Closing the Food Gap

Resetting the Table in the Land of Plenty

MARK WINNE

Beacon Press, Boston

Beacon Press
25 Beacon Street
Boston, Massachusetts 02108-2892
www.beacon.org

Beacon Press books
are published under the auspices of
the Unitarian Universalist Association of Congregations.

Printed in the United States of America

12 11 10 09 8 7 6 5 4 3 2 1

This book is printed on acid-free paper that meets the uncoated paper ANSI/NISO specifications for permanence as revised in 1992.

Text design and composition by Yvonne Tsang
at Wilsted & Taylor Publishing Services

Library of Congress Cataloging-in-Publication Data

Winne, Mark.
Closing the food gap : resetting the table in the land of plenty / Mark Winne.
p. cm.
ISBN 978-0-8070-4731-6
1. Poverty—United States—Prevention. 2. Hunger—United States—Prevention. 3. Economic assistance, Domestic—United States. I. Title.

HC110.P6W53 2008
363.80973—dc22 2007019704

THIS BOOK IS DEDICATED TO my mother and father, who taught me to care, and to those who came after them, who taught me to grow, cook, and taste food. It is dedicated as well to all my former English teachers, especially Mrs. Pipp and Professor Tagliabue, who taught me to express myself imaginatively, and to my children, Sarah and Peter, who allowed me to teach them.

And of course, *Closing the Food Gap* is dedicated to a thousand brothers and sisters across North America whose passion for food justice is no less than my own.

I am the people, humble, hungry, mean—
Hungry yet today despite the dream.

Langston Hughes

CONTENTS

PREFACE

As I WRITE THIS PREFACE to the paperback edition of *Closing the Food Gap* I'm struck by the similarities between food and energy conditions today and when I first started my community food career in the mid-1970s. Food prices are up by double-digit amounts. Energy, upon which our industrial food system is so desperately dependent, has reached never-before-seen prices. And for the poor, well, it's more of the same. There are now 28 million Americans receiving food stamps (the most since the program's inception in the 1960s), and food banks have been hit by unprecedented demand and food shortages. By all accounts the table in the land of plenty is looking a bit bare these days.

Back in the not-so-good old days of the 1970s we got our first taste of things to come. Food prices soared like a bull market on Wall Street, and fuel prices climbed so high that the nation's independent truck drivers went on strike because they were losing money. For a few days, grocery store shelves in some parts of the country were starting to look like those in Soviet Russia. We were learning a physics lesson the hard way: Food = energy. And within a few short years, the most conservative administration of the twentieth century took office, sending America's poor into food lines not seen since the Great Depression.

Even a grizzled food-system veteran like me finds no solace in watching history repeat itself. There has been too much waste, too much suffering, and too much exacted from our fragile environment for anyone but the most cynical observer to revel in the words "I told you so." Although I will confess to an occasional tone of smugness when I recollect the similarities between then and now, my real satisfaction comes in knowing that the ranks of community food activists I recount in *Closing the Food Gap* are swelling daily. Yes, the parallels between then and now are striking, but that so many people are joining the movement for a just and sustainable food system, and are eager to learn the lessons of the past, is profoundly moving. It gives me reason to hope that the America of the future will be a more promising place than at present.

As any first-time author soon discovers, the effort required to write a book is nothing in comparison to what is required to sell it. We slave away at our keyboards with an intrinsic faith that the reading public is just waiting to lap up our every word. Like naïve farmers who work from dawn to dusk to cultivate a righteous patch of produce, we soon learn that the stuff just isn't going to market itself. Perhaps in contrast to other "newbie" writers, I had the advantage of a great network of loyal comrades who have welcomed me into their communities to give talks, trainings, and lectures drawn from *Closing the Food Gap*. By the time this paperback edition appears on bookstore shelves, I will have visited nearly forty cities in states from Maine to Hawaii in less than one year.

At first these visits felt awkward. I was like some bumbling scribe mumbling a few lines of semi-literate prose. But when I realized that my listeners were indeed ready for hope and ideas, I began to feel more like an itinerant preacher spreading the gospel of food justice and sustainability. Pretty soon my sputtering road trip turned into a magical marketing tour. Wherever I went I was greeted by eager people who want to make good and affordable food a prevalent condition of community life—not just for the few but for all.

Farmers' markets, community-supported agriculture farms, and community gardens are exploding in number; food democracy is being cultivated by local organizations and food policy councils; and everywhere the drumbeat for economic and social justice is pounding. I've spoken at libraries in Kansas City and Phoenix, bookstores in Memphis and Portland (Maine), churches in Portland (Oregon) and Nashville, college campuses in Boston and Virginia, and food banks in Seattle and Hartford. I've written blogs and op-eds, and posted excerpts, all in the hope of getting the attention a no-name author desperately needs to sell his book. While I'm happy to report that the book is selling well, I am convinced that those who come to hear my talks and those who read my book are not doing so for the words alone but because they know that the time has come for real change in our food system.

What is the change they want? While some rail against the evils of the oligopolistic forces of multinational agribusiness, more of us have simply lost patience with the food gaps I describe in these pages. They have developed a keen intolerance for a place as wealthy as ours willing to accept 36 million hungry and food-insecure Americans. They refuse to accept the claims made by the fast- and processed-food industries that cheap, convenient food will set us free, when in fact it is robbing years of life from a generation of children. Their natural sense of fairness is challenged when they learn and see that too many people–mostly poor–have been abandoned by the retail food industry and left to fend for themselves in food deserts.

A realization is dawning that not only do these conditions represent a flagrant violation of twenty-first-century American values and standards, but they also represent a failure of our collective imagination and will. Based on the data generated annually by the U.S. Department of Agriculture, the percentage of Americans who are considered hungry or food insecure is almost exactly the same as it was when the data was first gathered in 1996. In spite of what amounts to one of the greatest sustained peacetime mobilizations in our history–the development and operation of over fifty thousand emergency food sites and a vigorous anti-hunger advocacy effort–we have made little progress in eliminating hunger. Yes, through a variety of means we have done much to reduce the food gap, but until we address its root causes–a fundamental social and economic inequity that results from poverty–we are fated to do no more than witness history's endless cycles. It is no longer enough to simply manage poverty through food programs; it is now time to end it.

One reason that I have not grown weary of the struggle is that I see a growing stream of young people coming into farming and community food work. They are as eager as I was at their age to change the world, but, fortunately, they are much smarter. They are flocking to college and university programs in sustainability, agriculture, nutrition, health, and related subjects to prepare themselves in ways I never thought possible. (When I requested a half-course credit during my undergraduate years for the volunteer work I did in my campus community to organize food programs, my sociology professor laughed at me.) I've been gratified to know that my book is being put to use in such programs; I've heard from many students and colleagues that *Closing the Food Gap* in its hardcover edition was assigned reading in their courses. Now that it is available in a more affordable, student-friendly paperback edition, I hope that even more young people will have an opportunity to get into it.

As I say at the end of the book, there is no lack of work for those who want to make our food system just and sustainable. And from what I've seen, there is no lack of those who are willing to do the work. With the depth and breadth of understanding that now exists on the subject of social and economic justice, we can perhaps work a little bit smarter and even look forward to the day when the food gap is a thing of the past.

MARK WINNE
August 2, 2008

INTRODUCTION

I've Come to . . . Shop?

To ENTER THE PARKING LOT of any Hartford, Connecticut, supermarket in 1979 required a sharp, reckless turn into a poorly marked curb cut. If you came at it too fast to avoid a collision with the suicidal driver heading right at you, you would bottom out your car's undercarriage on the lot's steeply graded entrance. Once in the lot, Hollywood car-chase skills were essential to maneuver across a parking area that was strewn with broken glass, overturned shopping carts, and potholes deep enough to conceal a bushel basket. Since the white lines marking parking spaces were faded or nonexistent, you left your car wherever it suited you.

Once you got inside the store, the first thing you noticed was the smell. It wasn't so much that "something has died" odor, but more the scent of something that rotted and was never fully cleaned up. When seasoned with a pinch of filth, marinated in gallons of heavily chlorinated disinfectant, and allowed to ferment over many years, the store released a heady aroma that brought tears to the eyes of men stronger than I.

Crunchy sounds emanated from the floor as your shoes crushed imperceptible bits of grit and unswept residue whose origins had long since been forgotten. The black and white floor tiles were discolored, unwaxed, and marred at irregular intervals by jagged brown stains that were forever one with the tiles.

Granted, these were pre–Whole Foods Market days. The supermarket industry did not yet have the technology that gives today's stores the soft, warm glow of a tastefully decorated living room. Instead, the humming neon bulbs, shielded by yellowed plastic coverings, cast a sickly pallor over the shoppers, the staff, and, worst of all, the food. The iceberg lettuce, already suffering from a 3,000-mile journey by truck, looked like the victims of a mass beheading. The rest of the produce case, from mushy apples to brown bananas, displayed a similar lack of life. A stroll down the meat aisle was as appealing as a slaughterhouse tour at the end of a busy day. Small pools of blood that had leaked from hamburger and chicken packages dotted the surfaces of the white enamel meat cases, the blood at times indistinguishable from the rust that discolored the chipped veneer. The atmosphere did not encourage a leisurely appreciation of food, nor did you feel like engaging in more intimate acts of product selection such as touching, squeezing, or sniffing. The fear of prolonging the unpleasantness made "grab and go" the prevailing modus operandi.

It didn't take too many trips to this sort of market before I was sufficiently motivated to go to a suburban grocery store. I was lucky; I owned a working automobile. Up to 60 percent of the residents in Hartford's low-income neighborhoods did not. Nor, as I would find out later, did the city's public transportation routes go to the suburban supermarkets.

My journey to the nearest full-size chain supermarket was six miles roundtrip. The store had easy vehicular access and a large, well-maintained parking lot, as well as shiny, clean aisles, floors, and food cases. The floor space available for product display was at least twice that of the largest remaining Hartford store, and the products were pleasantly arrayed. The produce section, though not brimming with abundance by today's standards, was quite ample and free of wilt, anemia, and other symptoms of imminent death. The store's staff was reasonably friendly, albeit prone to the lassitude common among those who must do repetitive, low-paying work. At least they would help me locate hard-to-find items; those requests were usually greeted with hostile stares by workers at the city stores.

Besides offering a generally more inspiring shopping environment, the suburban store had another point in its favor: it was cheaper. While not every item in the suburban store was priced lower than in the city stores, I soon found that I was probably spending 10 to 15 percent less for my weekly grocery shopping than I had been in Hartford. This proved to be true even for chains that still operated stores in both the city and the suburbs: the suburban unit had lower prices than its city cousin. How could this be? I wondered. The chain bought from the same wholesale suppliers, the stores had roughly the same pay and staffing structures, and they were only a few miles apart.

As it turned out, my revelations as a new resident of Hartford elicited not much more than a knowing sigh from colleagues and neighbors. The fact that city stores were inferior to suburban ones was nothing new to them. They had been watching the slow but steady abandonment of the city by supermarkets for ten years. "Yes," I was told on many occasions during my first year in the city, "the supermarkets have abandoned Hartford, and the poor, who can't get to the suburbs, pay more." "Supermarket abandonment" and "the poor pay more" became part of the lexicon of the organization I had come to lead, the Hartford Food System, and for many years to come, this prevailing understanding defined the food gap.

WELCOME TO HARTFORD, CONNECTICUT

I knew little of Hartford prior to my decision to take a job as the executive director of the Hartford Food System in 1979. Trips from my home in New Jersey to attend college in Maine a few years earlier had taken me along newly opened stretches of interstate highway that bypassed the city so efficiently that any thought of a stopover was strongly discouraged. I knew I had crossed a second river that wasn't the Hudson. The Tappan Zee Bridge's broad reach left no doubt that you had crossed that renowned waterway. But the highway bridge that led from Hartford to East Hartford over the Connecticut River had been designed to keep the river's existence a secret. Views of its gentle, tree-lined banks appeared to have been officially denied. After all, road engineers back then were not rewarded for their ability to incorporate natural features into highway designs but for their ability to pave the most territory for the least cost.

My first day of work in Hartford brought me into intimate contact with a city that had previously been placed off limits to the casual traveler. A car tour took me down long, wide boulevards intersected by neighborhood side streets that, on that chilly February day, were empty of all human activity except the occasional homeless man drawing life from a brown paper bag. Three-story brick buildings of late-nineteenth-century vintage gave many of the streets a graceful symmetry, although I noticed that several buildings had broken windows or plywood nailed over all the openings. Known as "perfect sixes," these buildings were evenly divided into six apartment units and were common both just north and just south of the city's downtown.

Crossing over I-84 from Hartford's Northend, I entered the city's downtown business district, still a fairly vital place in 1979. It was filled with bustling pedestrian traffic headed to and from corporate office buildings and signature sandstone department stores. The world famous insurance industry was a driving force, with buildings sporting the companies' tastefully understated logos: the Travelers' red umbrella, the Hartford's twelve-point stag (or *hart*, as in *hart ford*, the place along the river where the deer crossed), and Aetna's mountain.

Among the many things I would learn about the city's corporate culture was that Aetna was deferentially referred to as "Mother Aetna" for its oversized nurturing presence. The reality was that these insurance giants not only dominated the cityscape physically but controlled the city's financial, social, and cultural life as well. Though commanding in their presence, they eschewed the kind of entrepreneurial brashness of today's Donald Trumps and Steve Jobses in preference for a more button-down, paternalistic ethos befitting their high-toned Yankee origins. After all, they labored under the long-dead but still ironic gaze of Mark Twain, Wallace Stevens, and Harriet Beecher Stowe, who had burnished the city during its golden era.

Hartford's Southend had the feel of Boston's North End, with an abundance of Italian immigrants and foods and the kind of vibrancy that you wouldn't expect from the uptown actuarial crowd. Wood-frame houses built in the 1920s lined the streets, some of which were still graced with elegant shade trees, whose generous canopies would later provide a welcome respite from the summer's heat and humidity. In those months, old men leaning on canes would huddle together on street corners and in cafés holding fast to the customs of Italian village life, minus perhaps the public consumption of grappa. The non-Italian speaker had virtually no hope of finding his way into this closed brotherhood.

The older Italian families in the Southend, like the African Americans in the Northend and the Puerto Ricans in between, were now the predominant faces in a city that was moving rapidly from white to brown and black. Hartford's civil disturbances (aka race riots) of the late 1960s had signaled the demographic ascendance of the city's black and Hispanic communities and the steady exodus of its remaining white residents. Similarly, the movement of Puerto Ricans into the Southend—they had come to the area in the 1940s and 1950s to pick tobacco—provoked an uneasy reaction among the younger Italian families, whose trickle to the suburbs was becoming a steady stream.

The nearby suburbs of Wethersfield, West Hartford, and Bloomfield beckoned. Increasingly, they were viewed as safe havens for Hartford's middle-class residents, who were growing uneasy with the city's changing complexion and growing number of lower-income households. Housing prices in the suburbs were still relatively low, the shaded streets were virtually crime-free, and the commuter ride to Hartford's central business district was comparatively short. Perhaps most important, the suburbs' public schools were excellent, which is usually one of the important criteria for any young family in selecting a community to live in.

What my windshield tour of the city revealed on my first day of work was the early but undeniable signs of a city in decline. It was a place whose natural chemistry had worked tolerably well for a hundred years or more but had been thrown too rapidly out of balance now that circumstances were changing. Middle-class flight, aided in part by America's car culture and the emergence of the interstate highway system, left behind a giant sucking sound in urban cores across the country. Like the surf crashing against a beach then draining seaward again, middle-class families rode a tide of fear, disorientation, and resentment as they escaped to the suburbs. Sometimes with an undertone of racism and sometimes with an aching kind of liberal guilt, most white families, and later middle-class Hispanic and African American families, left the city when their resources permitted.

There was a strong personal dimension to these social and economic circumstances as well. As someone who had moved to Hartford to run a social change organization, I was expected not only to talk the talk but also to walk the walk. The moral burden that I had willingly accepted was that I should live in and be a part of the community that I was there to assist. To commute to Hartford from the suburbs—to be a nine-to-five do-gooder—was frowned upon and in some cases explicitly forbidden. To only *participate* in social action had a lower status than to be *committed* to the same. Think ham and eggs: the hen participates, but the pig is committed.

EXPLAINING THE FOOD GAP

This was the physical and psychological landscape that welcomed me, one that would generally deteriorate in the years to come. More important, it formed the backdrop to what was then and, unfortunately, remains to this day America's food gap. As in the case of supermarket abandonment of urban (and rural) areas, the food gap can be understood as a failure of our market economy to serve the basic human needs of those who are impoverished. But poverty contributes to this gap, creating a situation in which a person or household simply doesn't have enough money to purchase a sufficient supply of nutritious food.

Hunger—the painful sensation that someone feels on a regular basis due to lack of food—is a relatively rare phenomenon in America today, but it nevertheless afflicts a small number of U.S. residents on an intermittent basis. The more common form of food insufficiency is known as food insecurity, a condition experienced by a much larger number of people who regularly run out of food or simply don't know where their next meal will come from. As part of the annual census update, the U.S. Department of Agriculture conducts a survey that determines the number of people who are food insecure (generally between 10 and 12 percent of the U.S. population) and severely food insecure (3 to 4 percent of the population, until 2006 labeled "food insecure with hunger").

As our knowledge of the connection between diet and health has increased, the food gap has taken on yet another dimension, one that, ironically, includes the overconsumption of food. By overconsumption we generally mean a combination of eating too much of the wrong thing and too little of the right thing. Overweight and obese Americans now make up more than 60 percent of the population. Because of their association with the nation's increased diabetes rate and other diet-related illnesses, obesity and overweight are conditions that threaten the public health in ways that generally surpass the effects of hunger and food insecurity. As such, they have become central components of this country's food gap.

Yet as we will see, hunger, food insecurity, poverty, and overweight/obesity often have overlapping associations and connections, and as with supermarket abandonment, the community or environmental context is just as important as the income of an individual household. What we now call "food deserts," for instance, are places with too few choices of healthy and affordable food, and are often oversaturated with unhealthy food outlets such as fast-food joints. People who live in or near food deserts tend to be poorer and have fewer healthy food options, which in turn contributes to their high overweight/obesity rates and diet-related illnesses such as diabetes.

Perhaps one of the most frustrating and perplexing features of the food gap is a certain relativistic quality that has wormed its way into our food system over the past ten years. Just as lower-income groups make some small gains in closing the food gap by, say, having access to new food stores in city neighborhoods or benefiting from a marginal improvement in the Food Stamp Program, higher-income groups leap ahead with an increase in their purchase of organic and locally produced food. In other words, as trends in consumption associated with lifestyle and health expand one class's universe of choice and perceived health benefits, a lower, less privileged class barely catches up to where the other class was in the last decade. The gap never decreases and indeed, as we will see, often increases.

In all the ways that we think about the food gap, we must think as well about the food system. In its simplest sense, food system thinking doesn't permit us to isolate one segment of food activity from another. We can't, for instance, think only about farming without also thinking about eating. We can't set a price for a food product without being sure that enough people want it badly enough to pay that price. All parts of the system, from seed to table, are connected in a vast and complicated web, and the more we understand those connections, the more likely we are to narrow the food gap.

POVERTY VERSUS HUNGER

Hunger, food insecurity, and poverty present us with a chicken and egg proposition. Can we significantly mitigate or even eliminate the first two if we eradicate the latter? Or, if the latter can never be eradicated (that is, as Jesus said, the poor will always be with us), should we focus society's resources on hunger mitigation as the most humane and practical strategy? The manner in which we debate this question has consequences for how society chooses to close the food gap. While the failure of supermarkets to adequately serve lower-income communities represents a failure of the marketplace, the marketplace is functioning rationally (as economists would say) by going to where the money is. In short, if communities weren't poor, they would have supermarkets and, as we will see, the best and healthiest food available. To move forward in our understanding of the food gap, we must also understand the role that poverty has played in giving hunger and food insecurity such a firm foothold in the United States. And we must understand as well why we have chosen to respond to poverty and hunger in the ways that we have.

As an up-by-the-bootstraps kind of people, Americans have always struck an uneasy balance between poverty and the social welfare programs that have attempted to address it. In fact, many antihunger and antipoverty advocates assert that the public and private charitable sectors have never made a concerted and meaningful effort to eradicate domestic poverty. It is notable, in that regard, that in the course of reforming the country's welfare system, President Bill Clinton said we were ending *welfare*, not poverty, as we knew it. With the exception of an occasional burst of rhetorical and political fervor, such as President Lyndon Johnson's War on Poverty during the 1960s, our nation's approach to poverty has been to manage it, not to end it. And perhaps the best examples of good poverty management practices can be found in America's antihunger programs.

Of all the consequences that we ordinarily associate with poverty—crime; self-destructive behaviors such as drug and alcohol addiction; mental illness; shorter life spans; infant mortality; asthma; obesity and diet-related health problems; a failure to live up to one's potential in school, the workplace, or the community—hunger and severe malnutrition have traditionally given society the greatest pause. Whether as a result of fundamental religious teachings or innate human compassion, most of us will do what we can to prevent a fellow human being from teetering too close to the brink of starvation. Food is the basic human necessity in which we invest the most energy to produce, and it unites the human race in a universal spirit of awareness, sharing, and charity.

Although moral ambiguity and the failure of our political will stifle our ability to attack poverty with anything approaching a meaningful resolve, as human beings we are simply not wired to stand by and allow poverty's worst manifestation, hunger, to be ignored. It is for this reason that the United States has created a vast and complicated system of private and public antihunger programs—that is, poverty management strategies—that do not have an equal or obvious parallel anywhere else in the developed world. While European nations of social democratic leanings take a more aggressive approach to poverty than does the United States, their social welfare systems do not include the enormous and separate food assistance programs that have evolved in this country over the past seventy-five years. The federal government's fifteen separate food assistance programs, which collectively spent $53 billion in 2006 on food for lower-income Americans, are also collectively one of America's largest welfare programs. Though not generally regarded as adequate in terms of benefits to individual households to ensure their food security, they are without a doubt the most significant protection against hunger and food insecurity available to lower-income Americans. And as nutrition programs—even though sometimes their nutritional impact is marginal, if not occasionally harmful—they constitute a critical health promotion and disease prevention strategy as well.

But the evolution of these food programs is a testament to our political uncertainty over who should benefit and to what degree. As Dr. Katherine Clancy wrote in her essay "Sustainable Agriculture and Domestic Hunger," "The Great Depression spawned major welfare programs, including farm price supports, food stamps, and Social Security, and minor programs such as commodity distribution." The early history of food-related welfare programs and their evolution has received excellent treatment from Dr. Janet Poppendieck, who, in her book *Breadlines Knee Deep in Wheat*, traces the development of the Agricultural Adjustment Act in the 1930s and the Federal Surplus Relief Corporation. The inherent irony of these programs—the nation's first ongoing form of federal food assistance—is that they were actually designed not so much for hunger relief purposes but as "a convenient outlet for products acquired in efforts to increase the income of commercial farmers." What the New Dealers were doing apparently was establishing food relief programs to prop up agricultural prices. When the food relief side of the equation leaned too heavily in favor of hungry people, groups such as the American Farm Bureau and the National Association of Manufacturers intervened to ensure that free food did not reduce the demand for commercial food. The effect, calculates Poppendieck, is that individuals in 1936 received about five pounds of food per month, which was about 5 percent of what they needed.

According to Clancy's history of food assistance programs, a precursor to the modern Food Stamp Program limped along until 1943, when it was suspended during World War II. Farm interests led by southern Democrats kept the lid on the distribution of surplus farm commodities to needy people. Even as surpluses continued to mount following the introduction of chemical fertilizers and pesticides after the war, members of

Congress from urban areas could not push through permanent food stamp legislation in the 1940s and 1950s. It wasn't until 1961 that President John F. Kennedy created the current Food Stamp Program, albeit very modest in size, with one of the first executive orders of his administration. His action on food stamps was as much the fulfillment of a campaign promise to meet the needs of poor people as it was a means to placate senators from farm states, where concern over continued farm surpluses and falling prices was growing. The program was expanded significantly later in the 1960s when Senators Robert F. Kennedy and Joseph Clark "discovered" hunger in America.

After the Food Stamp Program, the second-largest publicly supported attempt to reduce hunger and promote nutritionally adequate diets was the National School Lunch Program, established by President Harry S. Truman in 1948. Today, National School Lunch also includes other child nutrition programs, such as School Breakfast, Afterschool Snacks, and summer meals in parks and at recreation centers, all of which provide free or reduced-price meals to lower-income school-age children. Widely credited by educators for its ability to help focus a child's attention on his or her schoolwork ("a hungry child can't learn"), school meal programs have proven essential to the psychological and developmental well-being of young people.

But like the Food Stamp Program, the origin of child nutrition programs served other political agendas. During World War II, a significant number of men were rejected for military service because they could not pass the standard physical exam. Much of the blame for the high rejection rate was ultimately laid at the feet of poor nutrition. That so many young men had such substandard diets that they were unfit for military service was a matter of national chagrin and a threat to national security. This was the impetus for the creation of the national meal program to feed malnourished children and thus to ensure that the nation's future soldiers were fit to fight its battles.

Now, we can be reasonably well assured that New Deal architect Franklin D. Roosevelt, John F. Kennedy, and Harry S. Truman were as compassionate as other Americans of their times. We have no reason to believe that they didn't want to ensure that every American was well-fed. Yet as astute and practical politicians, they realized that neither Congress nor the American people would support multimillion-dollar (soon to become multibillion-dollar) programs whose purpose it was to mitigate the poor health outcomes associated with hunger. So they found it necessary to dress these new social programs up in clothes that could be worn comfortably in both agricultural and national defense circles.

As time went on, liberal social programs continued to expand in scope and size. To the food stamp and school meal programs was added the Women, Infants, and Children (WIC) Program in the 1970s as a way to further assist the youngest and most vulnerable members of society with specialized methods of food assistance. The overt connections to agriculture, such as distributing food in commodity-size cans of peanut butter, sacks of cornmeal, or blocks of cheese, were disappearing (although the political alliance that had been forged between antihunger interests and large-scale agriculture was still intact). These awkward if not ludicrous forms of hunger relief were being replaced by sophisticated coupon and voucher systems that enabled their recipients to buy food in grocery stores, more or less like everybody else.

A major liberalization in the Food Stamp Program occurred in 1977 when Congress did away with the cash-purchase requirement. In the opinion of Zy Weinberg, a thirty-five-year veteran of antipoverty and antihunger programs, the cash-purchase requirement was one of the more odious features of the still fledgling hunger relief effort. Devised by conservative politicians, who grudgingly supported the Food Stamp Program but wanted to make it as difficult as possible for poor people to use food stamps, the program actually forced people to come up with cash—like a down payment on a house—to buy their food stamps. Weinberg credits the elimination of the cash-purchase requirement with bringing millions of needy people into the program.

Trained staff and efficient bureaucracies were emerging all across the country. Nutrition and food-budgeting education and training supplemented the distribution of food benefits, which increased their value in the lives of poor families. When combined with strong outreach work, record numbers of underfed Americans began enrolling in food programs that were substantially mitigating the most egregious impacts of poverty. Collectively, these and other social welfare programs, such as Aid to Families with Dependent Children (AFDC), were managing poverty reasonably well. At various times, there were even teams of government-funded legal aid lawyers available to help low-income families who may have been illegally denied food assistance benefits. While hunger and poor nutrition were still a regrettable part of many people's lives, America had succeeded in at least identifying the needy and was making a good faith effort to ensure that nobody went to bed hungry.

As a college freshman in 1968, I found myself profoundly moved by the images of starving children in the Biafra region of Nigeria—collateral damage, so to speak, of that country's civil war. I cannot recall having feelings as disturbing as the ones provoked by those images at any earlier time in my life. The suffering that those photos and a few magazine articles described literally sent shivers up my spine. I knew very little of the larger political or social context of the conflict, but the hurt of these people had somehow lodged itself inside me. It was so powerful that I was compelled to start a campuswide movement to raise money for Biafran famine relief.

Such feelings, I believe, are the same ones that compel most Americans not to ignore hunger. When properly stirred—as we sometimes are by the media, social activists, members of the clergy, or inspired political leaders—we respond to hunger with direct acts of charity or support of publicly run antihunger programs. Our understanding of the events that caused the hunger may be blurry. We may not care to delve too deeply into the sources of someone's suffering. And complex social, political, and economic explanations may soar over our heads or simply hold no interest. But when we can feel the hurt, we respond.

As you will see throughout this book, however, the bridge from empathy to the political will necessary to create profound institutional change is a wobbly one. As the food gap grows in both width and complexity, it continually calls upon us to develop a more sophisticated understanding of its nature so that we can cross more confidently from empathy to political action.

THE HISTORY

CHAPTER ONE

Suburbia, Environmentalism, and the Early Gurglings of the Food Movement

He acts it as life, before he apprehends it as truth.
Ralph Waldo Emerson

THE TWIN REALITIES OF poverty and hunger were not my personal experiences. Neither, for the most part, were they the experiences of a generation that would eventually embrace environmentalism, pioneer the "back to the land" movement, and plant the seeds for organic and local food. But for those like myself who had been as touched by the words of Dr. Martin Luther King Jr. as they had by those of Rachel Carson, our reality was shaped as much by our disquietude with the unraveling of the natural world as it was by the striving of disenfranchised people for social justice. As middle-class baby boomers, we were a generation of light, white, and bright young people largely free of economic hardship, physical toil, and a host of vulnerabilities that commonly befall people from less privileged circumstances. Yet our class and relative freedom from suffering and oppression did not mean that we were feckless. To the contrary, our sensibilities were constantly scoured by a society in conflict that did not square with our inchoate values.

It was not coincidental, for instance, that our instincts and ideals were often shaped in places that once had the attributes we most treasured. For me, Bergen County, New Jersey, was that place. In the 1950s, the Garden State's remaining truck farms and rolling hills cushioned the growth of tidy suburban towns just beginning to push beyond their pre–World War II boundaries. The soft spaces of a thousand suburban lawns were the stage upon which I performed life's early dances. Marinated in green, steeped in autumn foliage, and dipped in spring's robust fragrances, I was nothing more than a semipermeable membrane open to every one of nature's gifts. But when one of America's first shopping malls opened near the intersection of Routes 4 and 17, I felt the tremors of shifting tectonic plates.

My mother would pack my brother and me into the family Chrysler for a day of shopping at the mall. Though never one for car trips, I liked this one because it took us past a handful of vegetable farms that still lined the road for a portion of the ride. With my chubby face pressed against the car window, I was hypnotized by the soldier-straight rows of peppers, tomatoes, and cabbages that clicked rhythmically by. Something in me responded to the horticultural order that the farmer had imposed on the field. It was a marriage of precise lines, soft shapes, and green plants alternating with brown earth. Rolling the window down, I would close my eyes and inhale the field's full bouquet.

By the early 1960s, north Jersey was fast becoming the poster child for sprawl. The natural buffers that had defined our communities and fueled my imagination had been bulldozed into oblivion. The last twenty or so undeveloped acres that were still within my bicycle's range finally became the area's first multiscreen cinema complex. My mother's trips to the mall were growing more frequent because there were now more of them. And the farmland that had aroused every fiber of my ten-year-old being was now sprouting Lord & Taylor, Acme Auto Parts, Pizza Land, and enough parking spaces to land a bomber squadron.

By the time I was seventeen, I had traded in my Schwinn for a 1965 Chevy Corvair. This gave my girlfriend, Becky, and me a chance to escape the fast-food joints that were now erupting like boils along our town's commercial corridor. We would drive to an orchard that clung precariously to life near our exurban edge and be, as Dylan Thomas said, "young and easy under the apple boughs."

Some ten years later, I attempted to find the orchard where we had frolicked. All that was left were a few unpruned specimens kept as landscape accessories on the back side of two-acre lots that now sported million-dollar homes. A former high school friend who ran a landscaping business in the area told me that his customers often called on him to spray those apple trees in the spring with a chemical that inhibited blossoming, thus preventing the formation of apples. "They complain," he told me, "about the mess that apples make when they fall all over their manicured green lawns."

Admittedly, farming, gardening, and even a proximity to these activities were assiduously avoided in the well-tended suburbs of the 1950s and 1960s. Nature was only a concept, and its yucky reality should be held firmly in check. When it couldn't be avoided, the thinking went, just make sure it was well sanitized. Producing food for a living, like preparing meals from unprocessed, whole ingredients, was spurned. "We made sacrifices during the Great Depression and World War II," I heard many adults say, "but when it comes to food, we are now free from physical work and scarcity." But as I was scooping ground balls out of the lush green grass of my front yard and my father was driving golf balls down landscaped fairways, Rachel Carson was writing *Silent Spring*. Our tidy world was inching toward a showdown with the iron laws of environmental limits at the same time one woman was trying to warn us that we were about to crash into a brick wall. She whispered in our ears that we could not continue hell-bent down the road to Gomorrah without suffering a painful, if not fatal, accident. My parents' generation chose not to listen.

Looking back, I often speculate on just how my parents viewed the natural world. Although they were not religious people by any means, I'm convinced that when it came to nature, they took their cues from Genesis, believing that man should have dominion over all the earth. They fought crabgrass as fiercely as our nation was then fighting communism. Insects, regardless of their size, function, or level of threat, were obliterated with extreme prejudice. Employing the doctrine of overwhelming force, they sprayed our trees so relentlessly with chemicals that flocks of birds fell dead onto our perfectly clipped lawn. (Spring in our neighborhood was an intensely silent affair.) The roses, the only flowering plant on our embattled landscape with the courage to bloom, were so encrusted with white fungicide powder that many years passed before I knew they were red.

It wasn't until our family made friends with a neighboring Swiss-German couple, refugees from World War II, that I had my first encounter with a vegetable garden. Paul and Ilsa Wirth had been close enough to the deprivations of war to know the importance of food security and the need for home food production. They were kind enough to take me through their beautiful garden and give me a few basic horticultural lessons. At the age of ten, in the middle of a July Fourth picnic at their house, I wandered off to explore the neatly tended vegetable rows at the back of their property. Barely as tall as a mature tomato plant, I suddenly came face-to-face with my first vine-ripe tomato. Lush, full, and round, I reacted to the sight and scent of it with the same sense of awe that I would react with much later in life at the sight of my first naked female breast. My stroll that day through the Wirths' backyard garden opened my senses to a much larger world of possibilities.

Almost everybody who grew up in the 1950s and 1960s has a story to tell about bad food. The children of that era are uncompensated victims of the *Good Housekeeping* promise that canned, frozen, and prepackaged food would free humanity from hell's kitchen. If there is a generation in history that is entitled to bring a class-action lawsuit against the food industry that first invented this food and then manipulated our mothers into serving it, it is ours. At the trial such items as canned peas, Pop-Tarts, instant mashed potatoes, Campbell's soup, Frosted Flakes, TV dinners, and Kool-Aid would be entered into evidence. A jury of our peers would certainly find the defendants guilty after only ten minutes of deliberation, and the judge would pronounce the maximum sentence possible: the executives of those corporations would have to eat nothing but their own food for twenty-five years to life. And may God have mercy on their souls.

One of my earliest childhood food memories revolves around bread. I vividly recall my attempts to pry the pasty substance formed by chewing Wonder Bread off the roof of my mouth. Once mixed with saliva, the glutinous wad had a way of stubbornly adhering to the surfaces of the oral cavity. Although I don't recall thinking then that the experience was particularly disgusting—Wonder Bread was just another 1950s food product that everybody ate—I did form a lasting impression of what bread should be: gummy, white, odorless, and tasteless.

My college days and the dawn of the organic food era radically altered my position on bread, though not necessarily for the better. The first "natural" food store to open near my college campus sold a dark, rugged loaf that required two hands and a strong back to carry home. Not yet owning cutlery of sufficient temper and edge to slice such a brute, my dorm mates and I would claw our way through the crust with our fingernails to separate bite-size morsels from the larger mass. It was only out of self-righteousness that we chose to persist, discovering that if we chewed each mouthful thirty or forty times, it would be safe to swallow.

If food has a Middle Ages—a period when a dark curtain descended over its history—it certainly has to be the post–World War II era in the United States. It was the beginning of mankind's descent into industrialized food production practices that emulated the same assembly-line technology that Henry Ford applied to the automobile. In the case of grain production for bread, the industrial model meant that a couple of varieties of wheat seeds were developed for the attributes that were so treasured in Wonder Bread. During the course of establishing this extremely limited range of wheat varieties, hundreds of other traditional varieties that had been bred for their ability to adapt to local conditions soon disappeared from use. (My current home in northern New Mexico had more than two hundred different wheat varieties in use before World War II.) Today, harvesting the wheat on large farms of thousands of acres each requires an immense investment in farm equipment. A single combine, for instance, starts at $250,000. The grain is held in elevators until it is shipped via a multibillion-dollar rail network to mills and bakeries that produce uniformly perfect loaves for distribution by truck throughout America's interstate highway system. Human hands never touch the grain, flour, or dough. Cheap energy, federal wheat subsidies, and massive corporate and public investment in infrastructure (mills, highways, ports, and rails) make this system possible. It is the industrial model of agriculture and food production at its zenith.

Developments like these did produce a cheap loaf of bread. And unlike the early disciples of an alternative way—such as college hippies gnawing at unyielding crusts—most Americans were content with the perceived benefits that technology brought them. But societies and systems have a way of cultivating their own antithesis. The more the industrial food system evolved, the more pounds of agrichemicals we dumped on the earth, and the more our parents served us bland, processed meals, the more people sought a new path.

INTELLECTUAL FOREBEARS AND THE MERGING OF MOVEMENTS

Hunger, supermarket abandonment of urban America, and a growing discontent with food and the environment emerged as separate but equal problems in the 1960s and gradually swelled to a tidal wave by the 1980s. Each to varying degrees had its own antecedents, champions, and constituencies. They chose different projects and solutions, mobilized their respective constituencies, and enjoined political figures to intervene as they saw fit. But by the 1970s, a series of books had identified the failures of America's food system, provided an intellectual platform for action, and even suggested a rough synthesis of parallel but related movements. Books such as Michael Jacobson and Catherine Lerza's *Food for People, Not for Profit*, Joan Gussow's *Chicken Little, Tomato Sauce and Agriculture*, Richard Merrill's *Radical Agriculture*, and Frances Moore Lappé's *Diet for a Small Planet* provided thoughtful and analytical weight to the growing body of personal experience. They and many others set the stage for a broader and more comprehensive understanding of the issues that in turn would catalyze an infinite number of ideas and solutions, including the one I was to both follow and promote in Hartford.

Food for People, Not for Profit, published in 1975, was one of the first overviews of the U.S. food system's many gaps. The book came out as a complement to the first World Food Day in 1975. It was also hot on the heals of the Russian wheat deal, which had seriously reduced domestic supplies of U.S. wheat, and the nation's first serious energy price spike, both of which drove up food prices to never-before-seen heights. *Food for People* slammed corporate agribusiness and its bedfellow, government. Its enumeration of our health and diet sins was prescient in light of today's obesity epidemic. It included a 1974 article from the *New York Times Magazine* noting that "the major nutritional affliction in [the United States] is obesity," a stunning remark considering that obesity levels then were absolutely wimpy in comparison to those today. The authors accused major food corporations such as Coca-Cola and Kellogg of promoting "eating habits that squander limited food and economic resources and degrade already inadequate diets." They went on to say that "the American food crisis is also reflected in the prevalence of diet-related diseases, beginning with dental decay and often ending with a cardiovascular coup de grace."

Food for People singled out as culprits the decline of family farming and the rise of corporate farming, the human health hazards associated with the increased use of food additives and other chemicals, the loss of farmland to suburban sprawl, and the ascendancy of fast food and junk food in the American diet. Sound familiar? We were told how Procter & Gamble invested $5 million (that's 1975 dollars) to develop the packaging for Pringles, one of the food industry's most distinguished examples of unhealthy and overprocessed food. In 1973, Dr. D. Mark Hegsted of the Harvard School of Public Health said that "30 percent of the food products in grocery stores today could be thrown out and nobody would be the worse." Jim Hightower, author of the now famous *Hard Tomatoes, Hard Times*, reminded us about the collusion between our public universities and agribusinesses that produced those infamous bouncing tomatoes—the ones that now make us pine for a real, garden-ripe tomato.

One of the book's editors, Catherine Lerza, sounded a clarion call for the emerging organic farming movement: "The United States agricultural system's . . . ever-increasing dependence on energy-intensive agricultural technology[,] . . . the development of genetically engineered crops . . . and unecological farming practices fly in the face of every known 'law' of natural systems." Her statements about the overuse of pesticides, the early stages of factory livestock farming and feedlots, and the disregard for age-old biological farming practices warned us, in a way similar to Rachel Carson, of the highway wreck ahead.

The book brought global and domestic hunger into the open by presenting early hunger research and the failure of U.S. policy to make a difference. It is depressing to be reminded, for instance, that the United Nations categorized 460 million people as hungry in 1974 when we know that the same body has reported that more than 800 million people were hungry in 2007, in spite of repeated efforts over the past thirty-three years to reduce hunger. Perhaps the most disturbing parts of *Food for People* were the testimonies given to the Senate Select Committee on Nutrition and Human Need in 1970 by two U.S. Army captains, Terrence P. Goggin and Clifford Hendrix. The two West Point officers had been ordered to survey federal food programs to see if the hungry were being fed. After visiting fifteen counties in five states, Captain Goggin testified: "I was emotionally stunned in going from household to household seeing children staring at walls . . . because they weren't getting food. I was stunned by the experience of going off in a car to a shack where children, in my opinion, were literally dying, their minds were dying."

The hunger problem was best summed up in a 1974 report issued by the Senate Select Committee on Nutrition, cochaired by Senators George McGovern and Robert Dole, a bipartisan team who continue to work to end hunger and poor nutrition. The report stated that the fundamental issue facing the hungry "is not [so much] the mechanics of the food assistance programs as it is the fact of persistent poverty, and the continued tolerance in this country of a starkly inequitable distribution of income. In a nation . . . in which 40 million people remain poor or near poor, more than a food stamp or child-feeding program is at issue."

Food for People assembled the state of the nation's knowledge and understanding of food, hunger, and agriculture issues in one place. By juxtaposing the day's salient food trends—hunger and poverty, the decline of family farming, the failure of national farm policy, the increased use of pesticides and other toxic chemicals, the growth in non-nutritional food—the book presented a holistic picture of what was seriously wrong with the much-touted American food system. In the meantime, the American Farm Bureau was bragging to Congress that farmers were producing food so cheaply that by early February of each year, an average American household had earned enough money to buy its entire year's food supply. But in the nation's agricultural heartland, the volcanic load of chemical toxins was unraveling the fragile environmental balance that preserved wildlife as well as the health of the soil, air, and water. Maybe food was cheaper, but Americans everywhere were slowly getting fatter, sicker, and more accustomed to an unhealthy diet. And in Harlem, Peoria, Fresno, and countless other cities and hamlets in the richest country on earth, people simply didn't have enough to eat.

INSPIRED, ENERGETIC, NAIVE

Food for People, Not for Profit concluded with a long chapter on what people could do about America's food crisis. It presented a substantial inventory of organizations and initiatives that were already tackling one or more of the issues described in the book. As the counterculture became a little less oppositional, many in college, just out of college, or in their twenties were busily rolling up their sleeves and putting their sweat and money where their mouths had been. The Zeitgeist was one of action and projects designed to prepare a new world stew assembled from a host of alternative ingredients—whether anyone knew how to cook or not.

I first got my hands dirty in a small Maine city that had been crippled by the passing industrial age. It sat on the banks of the Androscoggin River, a once beautiful waterway whose paper and textile mills had turned it into a wide, open sewer whose contents were too thick to drink and too thin to plow. The people clung desperately to the few barely above-minimum-wage jobs that remained in the dying mills, shoe shops, and service industries.

Armed with nothing more than our own bravado and a certainty possessed exclusively by the young, some college pals and I set out to see if we could make a difference. We developed a community center that provided peer drug and alcohol counseling (our only "credentials," of course, being our own experiences). Imitating the Black Panthers, some of whose members were on trial at that time in New Haven, Connecticut, we started a breakfast program for the city's poor children.

We organized the Androscoggin Community Food Co-op, a food store owned and operated by its members for the purpose of providing good, low-cost food to local people. The co-op sold an eclectic assortment of goods at prices lower than those at area supermarkets. Its shelves were stocked with items such as canned Maine blueberries, Aroostook County potatoes, twelve-pound wheels of cheddar cheese, white bread from a local commercial bakery, and Mailhot Real-Link Sausages. The sausages were my favorite, coming as they did from a local business owned by two brothers who made the ground pork stuffing from an old French-Canadian recipe. The Mailhot brothers stuffed the sausage casings (long ropes of pig intestines) with a little machine that sprayed the mixture from a jet nozzle into the condom-like containers, twisted the long tubes at regular intervals to form the links, and delivered the tasty little devils right to the co-op's door.

As some kind of assistant manager of these enterprises, I started my days at six in the morning with a drive to a nearby egg farm. I opened up the church basement where breakfast was prepared and served by a couple of student cooks from a local vocational school. I prayed that they would show up on time each morning, hangovers permitting. Several elderly and very sweet French-Canadian ladies volunteered to keep the kids eating breakfast under control. The children would always show up hungry, but with enough energy to flick wads of donated USDA commodity peanut butter against the church walls. When the meal was over and calm had returned to the kitchen, I completed the paperwork that would secure a reimbursement from the USDA for about two-thirds of the breakfast program's cost. This government-imposed budgetary shortfall required that I then spend a few hours each week begging as much free food from local sources as I could.

By afternoon I moved on to the food co-op, where my duties included car trips to far-flung Maine outposts to procure (or beg for) food. Like everyone else in the early co-op movement, I developed tendinitis in my right elbow from slicing large wheels of cheese into half-pound chunks, a condition known among co-op groupies as "cheese elbow." I spent any spare moments on the phone rounding up co-op members to fulfill their volunteer requirements or to pick up the food they had ordered before we left for the day.

Those early co-op days demanded hard, uncompensated time but also, just as important, creative food procurement strategies. I recall one deal that involved a drive to the Maine coast, where I would buy lobsters at well below the market price from a fishermen's co-op, assuming that such acts of solidarity carried their own virtue. Having long ditched the impractical Corvair of my youth, I would drive the lobsters back to the city in my 1961 Chrysler, hoping that not too many would escape from their seaweed-lined crates and crawl around the back seat of the car. Since even at below-market prices the cost of lobster was still too high for the members of our co-op, I would deliver the lobsters to a local butcher, with whom I would exchange them for about one hundred pounds of hamburger. We would then sell one-pound containers of cheap ground beef to our co-op members. To this day, the economic benefits of this deal, if there were any, escape me.

In their own small and idiosyncratic ways, the breakfast program and the food co-op were emblematic of early efforts to close the food gap. They both bootstrapped their way to providing a set of services that were necessary and valuable to their respective clients or customers. The breakfast program fed children who would otherwise start the day with nothing in their stomachs, while the co-op saved the city's working-class families about 20 percent on staple food items. Yet they both relied on a good deal of volunteer or in-kind labor, which was never fully accounted for in determining the projects' actual operating costs. As long as there was someone there to do the work, those who needed the food received it. But when funding disappeared or was insufficient (as was the case with the USDA breakfast program) and/or the volunteers burned out, projects sputtered along until they expired completely, sometimes dying a slow, painful death.

ON TO HARTFORD WITH A PLAN

The food project organizing skills that I learned the hard way in Maine and would later hone for a few years in Massachusetts were about to be applied in Hartford. While an old mill city surrounded by pint-size rural towns and Boston's outer-ring suburbs were barely adequate training grounds, they had given me enough confidence to believe that I could take on a city that was moving briskly past its prime. Against the backdrop of Hartford's growing poverty and white flight, in 1979 I was handed the reins of the newly formed Hartford Food System. My job was simple: to increase the food self-reliance of the city's low-income families and improve the livelihoods of Connecticut's farmers. I would pursue this mission by organizing community gardens, farmers' markets, food co-ops, community supported agriculture (CSA) farms, food banks, food policy councils, new supermarkets, nutrition education programs, community canneries, solar greenhouses, new government-funded food and farm initiatives, and farmland preservation programs. In other words, my job was to develop a highly integrated local food system—one that had been outlined in *Food for People, Not for Profit* but never attempted in the real world. The task was daunting, but the good news was that the person who had written the book on the subject was also the person who had prepared a food action plan for Hartford: Catherine Lerza.

The process of putting the Hartford plan into operation was aided by a politically progressive administration that controlled city hall for a short period of time in the 1970s and early 1980s. Led by an urban reformer and visionary by the name of Nick Carbone, city government was actively and aggressively using all of its power and resources to develop new strategies to turn around Hartford's declining fortunes. Among these strategies was a set of initiatives that would place the city squarely in the role of economic and housing developer, with the hope of creating more jobs and better living opportunities. But what was unique—indeed, never attempted by any other U.S. municipality—was an initiative to address the city's food problems, specifically the increasing cost of food available to Hartford's lower-income residents. To devise that initiative, the city in 1977 had contracted with Catherine Lerza to prepare a plan that would serve as the foundation for the Hartford Food System and, in effect, all later attempts to create an integrated and comprehensive approach to local and regional food, nutrition, and agriculture needs.

Lerza's report, "A Strategy to Reduce the Cost of Food for Hartford Residents," was submitted to city government and a dozen or so community organizations in the winter of 1978. The report called "the inflation attacking Americans' budgets 'the new inflation' because it is centered primarily on the necessities of life, such as food and shelter, items that used to hold down the Consumer Price Index [the government's indicator of annual inflation], rather than push it up." Because the Northeast region of the United States is "at the extremity of the national transportation system, in one of the country's coldest regions, [food inflation is] particularly critical to low income people who comprise approximately two-thirds of the Hartford population." (Since the region is at the end of the country's food chain—assuming that California and other portions of the West and Midwest are somewhere near the beginning—the high energy costs associated with shipping food from those regions to New England increase food costs there by 6 to 10 percent.)

The impact of the "new inflation" on low-income families was severe. The report noted that the food inflationary growth and the lower incomes of Hartford's residents were compounded by the existing urban food delivery system, which provided lower-than-average quality at higher-than-average prices, factors that were exacerbated by supermarket abandonment of the urban core.

What was the answer? Lerza's strategy was to close the food gap with a set of four self-help approaches that, taken together, would constitute an alternative food system.

- Community gardening and youth gardening
- Solar greenhouses, cold frames, and rooftop production
- Food distribution projects such as food-buying clubs, co-op stores, co-op warehouses, and farmers' markets
- A food processing center that would include canning, freezing, drying, butchering, and storage

It was estimated that through a combination of these self-help approaches, the food bill of the average Hartford household could be reduced by $300 to $500 per year. The calculation and the development of this system rested on the assumption that these projects would be implemented in a linked and comprehensive manner. In other words, each neighborhood would have a couple of community gardens, a food-buying club, a farmers' market located within a one-mile radius, and ready access to a processing facility. The report acknowledged that the development of an alternative food system on a citywide basis, something that had not been attempted before, would take several years before it became fully operational and would require subsidized labor and a strong base of grassroots and community support.

The report also posited a set of outcomes from the implementation of this system. Improving food quality was chief among them and would be accomplished simply by producing, buying, and consuming more fresh produce. This improvement would be supported by nutrition education activities provided by a number of the organizations that were expected to work together. They included the city's public school system and the University of Connecticut Cooperative Extension Service.

In addition to the city's low-income residents, farmers were expected to benefit as well. The proposed food system would increase demand for local produce. This would simultaneously open new markets for farmers who operated within a fifty-mile radius of the city and increase their income. The growth in food marketing, processing, and production activities was expected to create additional farm and food distribution training and job opportunities for city residents. Lastly, the new farmers' market activity and the creation of urban gardens were expected to make surplus produce available to the single emergency food pantry that existed in Hartford in 1978.

JACK HALE: THE LONGEST-LIVING VETERAN

"In the 1970s, people thought a food system was a cool idea." That's how Jack Hale, executive director of the Knox Parks Foundation, Hartford's community gardening organization, remembered the beginning of the Hartford Food System. As the only person still working on food issues in Hartford today who also participated in the organization's founding, he qualifies as both founding father and the initiative's longest-living veteran.

I once saw a picture of Jack when he was in his twenties sitting astride a motorcycle with a full head of long hair and a beard. There was something impressive, even intimidating, about him in a rebellious, 1960s kind of way. Balding and packing a few extra pounds now that he's in his late fifties, he still projects an intellectual fierceness that balances his huggable, almost roly-poly appearance. He reminds me of the laughing Buddha: he can draw you in with his easy laughter and sweet nature, but then slap you upside the head with his wit and wisdom.

Jack got his start as a community organizer with an outfit called the Connecticut Citizen Action Group, which mobilized state residents in the 1970s around consumer issues. He moved on to run the Connecticut Public Interest Research Group (ConnPIRG), a state offshoot of Ralph Nader's Public Interest Research Group. I always thought it was the combination of organizer training and hanging around the blunt and brooding Nader that accounted for Jack's street savvy and sharp critiques. He always offered the kind of insights that I welcomed, for the way they elucidated troubling conflicts, but also feared because they'd blow away my most treasured assumptions.

As an avid community gardener, board member of Hartford's Down to Earth Food Co-op, and city resident, Jack had the credentials to make him a first-class urban food activist. His sensibilities and internal compass were oriented toward the emerging alternative food system, with its links to healthy food, environmental action, and old-fashioned New England self-reliance. But he also possessed the sophisticated intelligence necessary to navigate the city's complex multiethnic and multiracial settings, as well as the labyrinths of city and state government. He had the rare ability to walk into almost any situation and size it up correctly by asking a few simple questions.

Jack's practical skills and philosophical disposition placed him in a key leadership role. Ironically, none of the other dozen or so organizational members who established the Hartford Food System, with the exception of the Knox Parks Foundation, had any food, farming, or nutrition experience. In fact, the food system's initial movers and shakers were mostly men who not only couldn't fix a tractor or can tomatoes but also in all likelihood didn't perform household cooking or shopping duties either. While many were seasoned community organizers and social service administrators, only Jack knew how to reach out to farmers, plant a garden, and cut cheese. I might add that he was the only person from my new board of directors who went out of his way to welcome me and my family to Hartford—a nice touch that he accented with the gift of a fresh loaf of bread he had baked himself.

"The Lerza document definitely put meat on the bones of the Hartford Food System theory, and the rhetoric around self-reliance was as good as anything else," Jack acknowledged, "but clearly the city government's leadership was key, and the involvement of neighborhood organizations gave the effort authenticity." Indeed, the organizational model for the Hartford Food System was one of its most intriguing features. While the city provided direction and funding, community organizations such as South Arsenal Neighborhood Development (SAND), the San Juan Center, Southend Community Services, and the Community Renewal Team were the vehicles that would implement the Lerza strategy "on the street." The street knowledge they provided—who the neighborhood leaders were, how to navigate local politics, who could and couldn't be trusted—was combined with their specific program knowledge. Groups such as the Knox Parks Foundation, ConnPIRG, the University of Connecticut 4-H, and the City of Hartford Board of Education were given the responsibility of providing the technical know-how for community organizations and residents. This collaborative model was based on the assumption that the best way to implement the strategy was through a partnership that relied on the complementary assets of its member organizations.

The racial and philosophical subtext that undergirded the partnership was both a strength and a weakness. Jack pointed out, for instance, that Hartford's outmigration of the middle class had created acres and acres of vacant land, which many of the early Hartford Food System's visionaries saw as a tremendous opportunity for urban gardens. "The Knox Parks Foundation had gone into low-income, nonwhite neighborhoods that had lots of vacant lots with their well-intentioned, white, paternalistic strategy to start community gardens," he explained. "But there wasn't enough neighborhood buy-in. In some cases, there simply wasn't any neighborhood. Everybody had left, which is why many of these projects failed. People didn't seem to understand that the most important word in *community garden* is not *garden*. We [Knox] learned [to tell residents] that if there weren't at least ten neighborhood people at the first organizing meeting for a new community garden, we wouldn't show up. Every time we've weakened on that rule, we've gotten killed."

It was advice like this, which Jack gave me on a regular basis, that helped me find my way through the twists and turns of Hartford's racial politics and neighborhood idiosyncrasies. The advice was usually dispensed from a stained booth in a seedy pizza place near his office where we met for beers after work. I would use the occasion to vent about the boneheaded behavior of the many groups I was attempting to coordinate. Jack would simply laugh at my stories and then tell a little tale or parable of his own, which was usually punctuated with an aphorism that contained my lesson for the day. His insights were always original and his judgments sound. I was lucky to have him as a mentor.

The organizing theory of the Hartford Food System was that the racial disadvantage of essentially white-run organizations could be overcome, or at least compensated for, by largely black- and/or Hispanic-run organizations that would use their higher levels of trust in nonwhite neighborhoods to engage and organize low-income residents. This assumption made sense and would produce many good outcomes over the ensuing years. The philosophical and cultural differences between white and nonwhite organizations, however, presented a more formidable challenge.

The local food, "grow your own," organic, reduce-dependence-on-fossil-fuels-and-increase-self-reliance movement had its roots in environmentalism, the first Earth Day (1970), Rachel Carson, Henry David Thoreau, and Wendell Berry. Those were clearly the intellectual inspirations for the Hartford Food System and its "Strategy to Reduce the Cost of Food." Its primary proponents were well-educated and privileged white folks. They worked for the more technically oriented food organizations. By contrast, the neighborhood organizations that were attempting to interpret the strategy for low-income black and Hispanic residents were shaped by the civil rights movement, the War on Poverty, the Reverend Martin Luther King Jr., Malcolm X, and César Chávez. They considered social justice, an end to racial oppression, and neighborhood empowerment their guiding principles. Time and constant struggle would help these two powerful forces occasionally find common ground and even produce potent synergies, but their philosophical and historical tensions have remained a vigorous part of the debate, even in the present phase of lower-income communities' struggles to secure healthy and affordable food.

As Jack Hale pointed out, in spite of the racial and philosophical divide in the Hartford Food System partnership, the organization's purpose was to create an alternative food system that could serve everyone equitably and fairly, especially the poor. Unlike groups in other parts of the country that were organizing farmers' markets to help only farmers or creating food co-ops to help the college-educated organic food converts, the Hartford Food System accepted the unenviable task of trying to merge two separate but related movements. This was new, it was exciting, and it was risky as hell.

"I'D LIKE YOU TO MEET YOUR BOARD OF DIRECTORS AND STAFF"

My first day at the Hartford Food System was not a propitious one. In the middle of a February sleet storm, I got a flat tire on my way to work. I managed to pull my car over to the side of the interstate, where I changed the tire at a place that had been designated by the Federal Highway Administration as the most dangerous highway intersection in America. Later that day, several members of the board of directors—my bosses and representatives of the organizations identified earlier—graciously took me to lunch. But before we had finished our appetizers, a verbal round of fisticuffs erupted between two factions of the board. In what was a harbinger of the many turf battles to come, one group of organizations objected to another group's claim that they deserved the credit for the success of a recent endeavor. Water glasses hit the table with a thud, faces reddened, and angry hands bent the restaurant's flimsy utensils. Clearly, the harmonious and collaborative relationships that had been represented to me during my job interview had been exaggerated. I left lunch with a raging case of indigestion and wondered if it was too late to get my old job back.

Hartford's poverty rate was climbing steadily, while greed and grandiosity were turning Connecticut's countryside into a sea of McMansions. Given the apparent gravity of the situation, I was under the impression that I would have a crack staff of "food warriors" at my disposal. But like the cooperative spirit that was supposed to exist among the Hartford Food System's member organizations, the quality of our staff had also been overstated. A budget had been pieced together from a variety of sources, including city government, private foundations, and the U.S. Action Agency, which was the most recent incarnation of the Peace Corps and its domestic version, Volunteers in Service to America (VISTA). Since most of our staffing support would be drawn from Action/VISTA, the salaries would be at subsistence levels, and our ability to attract highly qualified people would be limited.

On one typical Monday morning during my first year of work, not a single member of my staff showed up. Among the reasons offered were the following: one was recuperating from a "bad case of nerves" she had experienced over the weekend; another, a former Black Panther who had been convicted of a felony in the infamous New Haven trials, was being questioned by the police; one had apparently been arrested the previous evening on robbery charges; one was recovering from an abortion; and the last staff member was seriously ill from a disease so rare that it afflicted only thirty people in the United States. The irony—which was truly tragic, in some cases—was that none of these excuses was a lie; they were all true. When asked by friends how many people worked for me, I generally replied, "Less than half."

I report these events not to mock or criticize my colleagues, but only to highlight how inadequate, unprepared, and underresourced we were for the task at hand. We confronted some of the most entrenched social and economic problems in the nation, yet we might as well have been trying to hit a Roger Clemens fastball with a Wiffle bat. I learned to play the hand that was dealt me, and over the course of many years, I was able to secure sufficient resources to upgrade the training and quality of the staff. As a result, good things began to happen.

The guiding assumption of the early days of the Hartford Food System, which wasn't so different from incipient efforts elsewhere, was that a merging of the environmental ethic with that of traditional antipoverty and neighborhood-based programs could provide enough momentum not only to lift the poor out of hunger but also to raise entire communities out of poverty. Although we would discover value in collaborations like these and make many worthwhile contributions to people's lives with projects such as farmers' markets and community gardens, it was hopelessly naive to assume that our impact would reach the exalted levels that we originally had in mind. And as we would shortly learn, a relatively liberal government climate that had begun to at least mitigate the worst of poverty's features would soon turn hostile and cold.

CHAPTER TWO

Reagan, Hunger, and the Rise of Food Banks

There is always a certain meanness
in the argument of conservatives.
Ralph Waldo Emerson

I MET ALICE M. IN 1983, just two years into the first administration of Ronald Reagan. She was seventy-nine years old, living alone on Hartford's Farmington Avenue, only a couple of blocks from Mark Twain's former mansion, where he wrote *Huckleberry Finn*. Her eyes still twinkled a clear blue in a face that was deeply wrinkled and rarely offered a smile. She received $256 a month in Social Security and an additional $32 a month in Supplemental Security Income (SSI), making her yearly income $3,456. This placed her significantly below that year's official federal poverty level of $4,680 for a single-person household.

Because she had a pacemaker and was generally nervous about being out in her high-crime neighborhood, Alice rarely traveled more than two blocks from her apartment building. Not able to drive, she did her food shopping by walking to a small nearby grocery store whose high prices and lack of bargain brands kept more mobile shoppers away. Since a nearby A&P Supermarket with better prices and more selection had recently closed, this store was Alice's only choice.

Her memory was spotty and selective, but she recalled that as a teenager, she would ride the trolley all over Hartford and into the adjoining towns, which at that time were mostly farms. At her mother's request, she would stop along the way at nearby farm stands to pick up some tomatoes or freshly picked corn. (The trolley is now long gone, the memory of its extensive and vital network barely recalled by the beautifully restored Trolley Barn, which used to house its cars but now houses young architects designing new suburban homes.)

Before 1983, Alice's meager food budget had been stretched by her allotment of $44 a month in food stamps. In October 1982, however, her food stamps had been cut to the minimum allotment of $10 a month as a result of a Reagan administration measure reducing food stamps when Social Security payments increased. (In 2007, $10 was still the minimum amount of food stamps that many elderly people were entitled to.) The first month that Alice's food stamps were cut, she was so mad she sent them back.

"Many elderly people receive benefits of only $10 a month," said Bill Brady, who ran Hartford's Division on Aging at that time. "For that small bit of help, they have to go through what many of them see as a humiliating, even threatening process, perhaps waiting in line for several hours, submitting to redeterminations [to make sure they hadn't won the lottery in the last month] every six months, and so on."

Alice's financial situation was not unusual for many of the city's elderly residents, 16 percent of whom lived below the poverty level. Grocery shopping has always been problematic for the elderly, even if they own an automobile and are in reasonably good health. Aging and the physical challenges attendant to it make moving around in crowded stores and handling bulky food packages difficult. But in Hartford, as in other cities, the problems were compounded by the lack of car ownership, limited transportation options, and, most of all, the absence of nearby, high-quality, affordable food stores. The city offered the Dial-a-Ride transportation service for shopping and medical errands, but it was so rigidly scheduled that it was inconvenient for many elderly citizens.

Helen S., age seventy-eight, was another low-income Hartford resident whose combined Social Security and SSI totaled a miserly $292 per month in 1983. Her rent was $140, not including heat; her heating bills averaged $115 a month during the winter. Helen was diabetic and had high blood pressure, so she also had high medical expenses. Because of all these factors, she had received an increase in her monthly food stamp allotment from $54 to $68. But then Ronald Reagan took office, and Helen's food stamp allotment was reduced by about $21 per month.

The Reagan food stamps cuts, totaling $1 billion nationwide, were created by a host of minor changes in the formulas used to calculate food stamp eligibility and the amount of food stamps given to each person, in what might be called death by a thousand paper cuts. In addition to cutting food stamps, the Reagan administration was looking for every stone from which it could squeeze blood. One of its cruelest cuts was to congregate meals, which were served in community and senior centers five days a week. Though technically available to anyone over the age of sixty, senior meals programs tended to be used more by low-income elderly people than by others, both for financial reasons and because of the centers' proximity to lower-income communities. When Reagan took office," the Community Renewal Team, Hartford's community action agency, was serving 380,000 meals per month at 28 different sites. His cuts effectively reduced those meals by 30,000 per month.

Food, housing, and medical expenses tend to be the "big three" cost factors in the lives of lower-income people. A shortage of low-cost housing drives up rents, just as a shortage of affordable food stores in lower-income areas drives up the cost of food. Medicare and Medicaid tend to keep the lid on health care costs for the elderly and low-income households, but the coverage is never enough to fully protect the uninsured and those who experience major health problems. Like anyone else, lower-income households are at the mercy of the market when it comes to finding affordable food, shelter, and health care. Federal subsidies, along with the interventions of private charitable organizations, are designed to close the gap between the marketplace and the actual capacity of people to pay for goods and services. But as the marketplace changes, which generally means making goods and services less accessible to the poor and often more expensive for everybody, and the support of the public sector waivers, which means never keeping up with the marketplace, the poor get less, pay more, or simply go without.

In the case of the elderly, the pressure of the big three is especially acute. In the early 1980s, in addition to the declining value of federal food assistance and what was then becoming a greater dependence on the charitable sector, the elderly had to cope with declining housing opportunities and unpredictable health issues. Juan O., a sixty-six-year-old resident of Hartford in 1982, typifies the bind that housing prices create for many low-income elderly. His income of $268 per month, which included a $10 food stamp allotment, had to cover all of his living expenses. His rent for a dilapidated apartment in the city's Southend, including heat and electricity, was $180, which left him with $88 for food, clothing, transportation, a share of his medical bills, and all other necessities. In spite of repeated efforts, Juan had not been able to get into subsidized senior housing or a state Medicaid program. The city's 16 subsidized elderly housing complexes had 1,900 apartments but a waiting list of 1,500 people.

Health care costs can be particularly devastating for the elderly. For instance, Alice M. was trying to pay off a $300 dental bill in monthly installments. Medicare didn't cover dental work, and she had decided not to apply for a state-funded Medicare program that did provide dental coverage because she would have had to turn over her small life insurance policy to the state. Welfare assistance programs have always had very stringent limitations on personal assets.

The situation facing the city's elderly was best summed up by one of Hartford's Division of Elderly Services workers: "We can hook them up to the programs that are available, but the benefits are too small." With the nation's fiscal conservatives sucking savings from every source and the marketplace yielding no ground to those living on limited incomes, the prospects for elderly people gaining any measure of financial security were rapidly fading.

THE SAFETY NET UNRAVELED

The fragile but critical role that federal food programs played in the lives of Americans became apparent within twelve months of Ronald Reagan taking office. In the face of a softening economy and growing unemployment, the Reagan administration moved swiftly to implement its conservative agenda by cutting the programs that poor people depended on for survival. The administration's political philosophy of fiscal conservatism and personal responsibility allowed for little money to end poverty and little interest in government interventions to reduce hunger or poverty.

In urban communities such as Hartford, where the outflow of middle-class families was accelerating from a trickle to a torrent, the growing concentration of lower-income families faced diminishing economic alternatives and declining schools, all of which placed a greater burden on already strained social services. With a "take no prisoners" conservative occupying the White House, a contracting national economy, and the once vibrant urban centers hollowing out, all the conditions were ripe for a perfect socioeconomic storm. The food safety net that had been diligently woven over the past twenty years was now being snipped one strand at a time. The pressure to feed a rising tide of hungry people fell increasingly to local communities and especially to nonprofit organizations and faith-based institutions.

But rather than rise up in rebellion against the mean-spirited ideologues who controlled the government—an action suggested by more than one activist of that period—communities found new reservoirs of charity and compassion with which to stem the growing tide of need. As if to say "We don't have time to organize effectively against this kind of wrong-headedness," grass-roots groups fell back on a kind of quintessential can-do American spirit to address the crisis at hand. Believing that the flow of hungry people could be stanched by securing and distributing unwanted food, thousands of paid workers and volunteers started soup kitchens, food pantries, and eventually the country's biggest charity, America's Second Harvest—The Nation's Food Bank Network. In response to a fraying federal safety net and with lines of hungry people snaking around the block, the emergency food movement was born.

Churches and private social agencies in Hartford were scrambling to feed the rapidly growing numbers of hungry people who started knocking on their doors in 1981. Food pantries and soup kitchens were springing up throughout the city, as they were in cities across the nation. But even in the first rush of excitement that comes when people rise as a team to confront a tough task, most people felt that they were not even coming close to filling the need that had been created by high unemployment, widespread poverty, and deep cuts in federal social programs.

"We are the private sector that Reagan says will take up the slack, and we're getting killed," said Lola Elliott-Hugh of the Salvation Army of Hartford. Between 1980 and 1982, demand on the Salvation Army's food pantry increased an astounding 400 percent, forcing a reluctant decision to stop serving single men, who seemed to be the least vulnerable. Decisions like this reflected both a stated and an unstated method of triage, whereby the service provider would decide who was more needy and occasionally who was more deserving. While these decisions were sometimes based on the most rational criteria that people could employ at the time, they also opened the floodgates to arbitrary, subjective, and hugely biased judgments made by highly prejudiced people. Without any standard or means test, or any other commonly accepted screening tool, volunteer organizations were setting themselves up as judge, jury, and executioner.

Although other increases in demand were less dramatic, all of the emergency food sites in Hartford reported more people seeking help. In addition to the numbers, which ranged from a low of 15 percent to the 400 percent increase reported by the Salvation Army, the other striking feature of this demand was the characteristics of the people seeking help. In addition to the usual street people, the sites were beginning to see more young families and unemployed young men. One church worker said that families that had previously been able to stretch their money until the last few days of the month were now showing up at the food pantry by the twentieth day of the month. Regrettably, the pantry could give each family only a two- to three-day supply of food, which left a gap of as much as a week before the family could get further food assistance.

The city's faith communities joined forces with their suburban counterparts to increase the supply of food and the number of volunteer hands to feed the hungry. One group of mostly suburban Protestant churches was providing a regular stream of donated food to the Horace Bushnell Church, located in one of the city's poorest neighborhoods. The area's Catholic churches banded together to assist the House of Bread, a soup kitchen operated by the Sisters of Saint Joseph. Even the newly opened Hartford Farmers' Market was making regular end-of-the-day produce donations. Churches and synagogues also were crossing denominational lines to meet the need, as nine downtown faith communities created a consortium to operate a food pantry, a soup kitchen, an elderly feeding site, and an emergency shelter.

Some groups ventured beyond immediate food relief and attempted to address some of the food emergencies that people experienced. One staff member at the Church of the Good Shepherd provided counseling, budgeting advice, information about legal rights, and even occasional interventions with government agencies that were now making it more difficult for people to get food assistance. To increase the overall food supply, Connecticut opened its first food bank warehouse in New Haven in 1981 to solicit, hold, and distribute larger quantities of food donated by food industry sources. And to encourage more food companies to donate food to the food bank and food pantries, advocates went to the Connecticut legislature to gain passage of the Good Samaritan Act, which reduced the potential liability of food donors.

Despite the increased activity around emergency feeding, activists held no illusions that enough was being done. They knew that the emergency food system as it was then evolving was not capable of handling the thousands of people who were asking for help. As people were deciding whether to eat or heat and federal cuts got closer to the bone, the embryonic emergency food system was starting to recognize its own limitations. "We are just scratching the surface," said Mark Patton of the newly formed Connecticut State Food Bank, "and we know that the food pantries don't have any reserves."

Even with the intervention of a large number of wealthy suburban churches and synagogues, Hartford's providers were running out of food. The Horace Bushnell food pantry frequently went dry. One food pantry that had not been open for very long was forced to close when it was unable to keep up with demand. And despite the enthusiasm of food pantry volunteers for their own projects, no one in the emergency food business sought publicity for their services, for fear of being overrun by needy clients. In addition to the nervousness over the growing numbers of hungry people, many faith-based activists expressed frustration about the role that had been forced upon them by the failure of the federal government.

Mark Patton summed up the Zeitgeist of the moment when he said, "We are playing into Reagan's hands by increasing private feeding activity while the federal government is doing all it can to shirk its responsibility. This patchwork system is an inadequate and terribly inefficient way to try to keep people from starving. But at the moment we have no choice." Little did he or anyone else in the ragtag emergency food movement know that this was just the beginning of what was to become a sprawling, multibillion-dollar network. More important, the organizers, who honestly believed that their actions were a temporary response necessitated by a true emergency, could not have foreseen that their work would permanently alter the way the general public—and government at all levels—would address the food gap.

The reemergence of hunger at this level, as a national phenomenon, stunned everybody. But in addition to the response it evoked, it also brought into focus ironies in the nation's food system and the gap between the haves and the have-nots. In the same year, 1982, that soup lines were longer than at anytime since the Great Depression, the federal government was stockpiling surplus cheese from dairy farmers faster than it could give it away. Grain harvests were so abundant that farmers were going out of business due to record low prices. And the American food manufacturing industry was pumping out at least 15,000 "new" food products each year, many of which had no nutritional value, would never sell well enough to remain on the shelves (and would then be donated to food banks), or were, as would later be discovered, downright harmful.

Many of the problems underlying the U.S. food and farm economy were not necessarily new. Farm surpluses had long been a problem among farmers who produced commodity crops such as grains, soybeans, and dairy products. Indeed, the overproduction of farm products had led to the manic creation and distribution of many non-nutritious food items. What was new at the time was the federal response to the surplus. In the 1960s and 1970s, the federal government had thought it wise to help poor people gain better access to our abundant national food supply (and, it should also be said, to help farmers gain access to additional markets). While federal programs such as food stamps, National School Lunch, WIC, and elderly meals didn't correct the underlying problem of poverty, they did have the straightforward virtue of keeping millions of Americans, especially children and elders, from going hungry. In the 1980s, the Reagan administration broke the contract with the poor and passed the buck back to civil society, without whom millions of people would have starved.

HUNGER COMES OUT OF THE CLOSET

Long lines curling along the streets near soup kitchens and food pantries were beginning to convince some Americans that the nation had a food (and a poverty) problem. A few community food activists yelled and screamed at their city councils, state legislatures, and the press, but their voices were muted by a collective incredulity that restrained action. Part of the problem may have been that the conservative political climate of the 1980s cast such a pall that government was cowering like a dog that had been whipped too much. But part of the problem also lay with those who would normally have been outraged by such a violation of the social contract. Instead of opposing the policies that were driving people to emergency food programs, they were focusing on building higher dikes to contain the rising tide of hunger.

As someone who was more comfortable getting his hands dirty with local food projects such as farmers' markets and community gardens than with tramping down the hallways of the Connecticut legislature, I joined the frenzy to expand the emergency feeding network. I suspended my rational thought processes and found it more gratifying to throw sacks of donated potatoes around than to spend months, often years, trying to secure incremental funding gains for food programs from the legislature.

In 1982, as part of my work at the Hartford Food System, I became a cofounder and first board chairman of the Greater Hartford Foodshare Commission (called Foodshare today), the region's food bank. Developing a new nonprofit organization requires a big investment of time and energy and can easily distract one's attention from potentially more important matters. The prospect of the food bank, with all of its volunteer energy focused on securing and distributing donated food, was exciting. It quickly enveloped all of us in a halo of good karma. But there was an opportunity cost associated with choosing to collect and distribute other people's leftovers rather than fight the public policy battles that should have been fought more vigorously. The result was that the public lost sight of the need for fundamental change, and policymakers were let off the hook. The goodhearted donors whom we had enlisted in the battle to end hunger thought that they were doing the right thing.

Fortunately, the thrill of dishing out tuna casserole at a soup kitchen began to dissipate, which may be why attitudes began slowly to shift by the mid- to late 1980s. There were also signs of fatigue in the emergency food world. The fervor that had fueled the food pantry crusade to end hunger was beginning to abate. The realization that a political solution to the war on hunger was necessary was beginning to creep into our consciousness. One intriguing reason why the shift began to occur was the situation in Africa, never a stranger to hunger. Through a set of economic and natural circumstances, almost twenty-five African nations were once again teetering on the brink of famine. To gather some much-needed resources and draw America's attention to Africa's plight, performers Michael Jackson and Harry Belafonte and music promoter Ken Kragen assembled an armada of pop stars to make the album *We Are the World—USA for Africa*. A great success as both a fundraiser and awareness builder, the enterprising effort sent millions of dollars to starving Africans. Just as important, America's attention was once again focused on hunger.

Within a year of the album's release, the same cast of characters, led by Kragen, hatched an idea to focus attention on U.S. hunger, which they called Hands Across America. Their plan was to create an unbroken chain of people holding hands from America's east coast to its west coast. The participants sought sponsorship from friends, relatives, and neighbors. More a cool idea than a logistical possibility, Hands Across America had several major holes in its "chain" on the day of the big event. However, it did succeed in raising several million dollars, which it would distribute to hundreds of food organizations, most of them emergency food agencies, across the country.

For some reason, I had been picked by Kragen's people to represent Connecticut at a meeting in Los Angeles where we would decide how the money was to be divided. Based on Connecticut's participation in the national hand-holding event, its poverty data, and some advanced social calculus, the state would receive $76,000. We had two choices of how to use this princely sum: either spend all of it on one project or divide it evenly among the fifteen food pantries that had formed the Connecticut link in the chain. I was able to secure an agreement from all those involved to invest the entire amount in a new organization called the Connecticut Anti-Hunger Coalition (CAHC). This nonprofit organization was created in 1986 to advocate on behalf of the poor for more antihunger support from the State of Connecticut. In light of the fact that so much energy had been expended over the past five years to develop Band-Aid services, I considered the agreement to be a milestone. Those fifteen emergency food pantries had each given up a few thousand dollars, enough perhaps to feed their respective clients for a month. In return for this small sacrifice, we developed an organization that would, over time, leverage millions of dollars in state support for a vast array of antihunger and healthy food initiatives.

The funding enabled the CAHC to hire a full-time professional director, increase the skills of hundreds of individuals to secure beneficial changes in public policy, and educate the general public about the scope and depth of hunger in the state. It was the last function—public education and awareness—that proved to be the most valuable over time. This would become evident when Connecticut became the testing ground for the first comprehensive national project to document hunger using rigorous research methodologies. The CAHC and a Hartford organization known as the Hispanic Health Council would do what long soup kitchen lines and fervent pleas from activists had failed to do: make the prevalence of hunger in the 1980s a reality that couldn't be denied.

HUNGER REVEALED, HUNGER DEFINED

Many attacks on the nation's conservative political agenda had been made by advocates for the hungry in the 1980s, but few had successfully penetrated the Reagan administration's armor. It was felt, however, that if hunger could be both documented and defined, especially with a level of scientific rigor that was often lacking in the arguments of everyday advocates and food bank operators, the public and policymakers would pay attention.

To that end, advocates and academics joined forces to develop a research tool known as the Community Childhood Hunger Identification Project (CCHIP). Developed by the Connecticut Association for Human Services and sponsored by the Food Research and Action Center (FRAC) in Washington, D.C., the project was undertaken in nine U.S. cities, including Hartford, in 1989. The Hartford CCHIP was managed by the Hispanic Health Council and involved in-depth surveys of 315 households selected for the study because their incomes were no more than 185 percent of the federal poverty level ($23,495 for a family of four in 1989). Households were asked eight questions about their eating habits and choices in the previous twelve months as well as the past thirty days. The questions were as follows:

1. "Did your household ever run out of money to buy food to make a meal?"
2. "Did you or other adult members of your household eat less than you felt you should because there was not enough money to buy food?"
3. "Did you or other adult members of your household ever cut the size of your meals or skip meals because there was not enough money for food?"
4. "Did you ever cut the size of your children's meals or did they ever skip meals because there was not enough money for food?"
5. "Did children in your household ever eat less than you felt they should because there was not enough money for food?"

6. "Did children in your household ever say they were hungry because there was not enough food or money for food?"
7. Did you ever rely on a limited number of foods to feed your children because you were running out of money for food?"
8. "Did children in your household ever go to bed hungry because there was not enough money to buy food?"

Families that responded affirmatively to five of these questions were classified as experiencing hunger. Families that responded yes to two of the questions were considered to be at risk of hunger. And families that replied no to all eight questions were categorized as not hungry.

The Hartford study found that 41 percent of the families interviewed experienced hunger and 35 percent were at risk of hunger. This meant that an astounding 76 percent of the city's families that lived at or below 185 percent of the federal poverty level had significant food problems as a result of constrained resources.

What did these findings mean in terms of actual numbers of people and the impact of hunger on their lives? Between 30,000 and 40,000 Hartford residents, 22 to 31 percent of the city's population, were struggling to feed themselves and/or members of their families. A number of health and behavioral problems also were noted among those who were at risk of hunger. Hungry children, for instance, suffered twice as many health problems as non-hungry children, including significantly more episodes of fatigue, dizziness, headache, and inability to concentrate. Hungry children also were three times more likely to be absent from school than non-hungry children.

In terms of behaviors that hungry people employ to avoid running out of food (or to ensure that they have enough food), a new application of an old phrase crept into the antihunger lexicon: *coping behaviors*. Sounding more like the way one deals with an annoying mother-in-law than with securing one's sustenance, the term *coping behaviors* refers to a range of activities—from socially acceptable to downright criminal—that someone might use to get food or the money to buy food. These include stealing, dumpster diving, borrowing money for food, getting food from friends or relatives, sending children to someone else's house to eat, buying food on credit (common at many of Hartford's smaller grocery stores), and going to a soup kitchen or food pantry. The CCHIP study found that hungry people resorted to these behaviors significantly more frequently than non-hungry people did. For instance, hungry people sent their children to a neighbor's or friend's house to eat seven times more often than non-hungry people did, borrowed money three times more often, and used food pantries three times more often.

As to the ultimate causes of hunger in Hartford, the report's authors didn't mince words. They made it clear in their analysis and later in a set of bold policy recommendations that the lack of income (poverty and near-poverty), the insufficiency and ineffectiveness of major food assistance programs such as food stamps, and inadequate access to affordable retail stores were too blame. Pointing out that 77 percent of those surveyed who were hungry had incomes below the federal poverty level, the report strongly recommended that local, state, and federal government must do more to increase the benefits of food assistance programs, as well as outreach efforts to increase participation in those programs. With respect to limited food access, 20 percent of the respondents said that they were unable to shop at their first-choice food store because it was too far away and they did not have an automobile.

In addition to recommending significant increases in food assistance spending, the CCHIP study recommended action to increase the number of nearby, high-quality, affordable food stores. They pushed as well for more community gardens, farmers' markets, and direct-purchasing programs between the region's farmers and local institutions such as schools. In other words, one of the first and most thorough attempts to document the existence of hunger and promote a better understanding of its causes did not prescribe more and bigger emergency food programs. Instead, the researchers walked gently into the lives of the poor and respectfully asked them to share their stories. They found that these families' poverty, the systems that were supposed to help them manage that poverty, and the failure of the marketplace to serve their food needs were restricting their ability to live normal and healthy lives. The reactions to these findings would be swift and, over time, profound.

A steep learning curve and a roller-coaster ride of emotions marked my first ten years in Hartford. My feelings ran the gamut, from the excitement of self-help projects such as gardens and farmers' markets fueled by the hope of self-reliance, to the brutal realization that a mean-spirited political climate was the source of so much despair in people's lives. The rush of community mobilization to build a network of food banks that could relieve the suffering but never touch the root cause would be matched with equal intensity by the rush of advocates and academics to "prove" the damage that bad policies had caused. It was a whirlwind of activity and a cauldron of frustration and anger, but no one could have asked for a better classroom than the one given to me in Hartford. At the very least, I was well prepared for the challenges that followed.

THE REACTIONS

CHAPTER THREE

Farmers' Markets

Bringing Food to the People

"What's your favorite vegetable?"
asked the *Hartford Courant* reporter.
"Pork chops," replied the farmer.

THAT'S HOW THE OPENING EXCHANGE went between the press and a Hartford area farmer on the first day of the first farmers' market in Connecticut since World War II. That the market's organizer, Sally Taylor, had not rehearsed these pioneer farmers for their public debut was not her fault. The simple fact that nearly twenty Connecticut River Valley growers had circled their hodgepodge assortment of trucks and vans around the city's Old State House was a miracle. It had required months of detailed planning and delicate negotiations with skeptical city officials. Hardest of all had been the careful and extensive outreach to the region's farmers, whose initial response to the idea of an urban farmers' market ranged from a monosyllabic grunt to stone silence.

The chaotic street scene that unfolded on that July day in 1978 belied the deliberate behind-the-scenes orchestration that had been necessary to bring this dream to fruition. Hartford residents and downtown office workers had been forewarned with leaflets and press releases, but no one had quite imagined the disheveled band of country folk who would appear on their doorstep to hawk their wares. The farmers were backing their trucks up to the curbs in their preassigned spots, just as planned. Signs and banners had been dutifully placed to secure maximum exposure but avoid offending the city's insurance executives. The city's police officers, with their squad cars' cherry tops spinning, were there to direct traffic and protect the public safety, just in case the pent-up passion for fresh vegetables erupted into violence. No laughing matter, the negotiations with the Hartford Police Department had been among the most difficult and had resulted in a pledge that the event's organizers follow every rule to the letter or be prepared to give up their first born child.

The presence of the local constabulary heightened the drama. Is someone famous coming to town? Is there an office tower fire? Where's the bank robbery? These were just some of the questions coming from would-be customers starting to close in on the scene. With a sense of heightened expectation, the number of people began to swell, seduced perhaps by the heavy scent of fresh peaches, just-picked corn, and brilliantly colored flowers. People who had rarely, or in some cases never, seen, smelled, or tasted local produce found themselves caught up in the gawking crowd. They would soon engulf the hapless farmers, who took to defending their rickety card tables like soldiers on an embattled perimeter. As a group that generally avoided Hartford at all costs, the farmers found their pockets swelling with cash in direct proportion to their dwindling supply of produce.

Farmers' markets started popping up here and there across America in the mid-1970s. Los Angeles, Santa Fe, and New York City were among the early entries as consumers began to rediscover farmers as well as the taste and price value of locally produced food. For Hartford, farmers' markets were a part of the original Hartford Food System game plan. The idea was simple: connect local farmers and urban consumers, especially lower-income families, for both groups' mutual benefit. This could be done by developing higher-return retail outlets for small and medium-size farmers, while providing more high-quality outlets for produce-starved urban consumers. In Hartford's case, unlike that in many other cities that were developing farmers' markets at the time, meeting the needs of low-income city residents was just as important as meeting the needs of farmers.

The assumption was that the increased demand for local produce, which could be generated by the proposed food system, would stabilize the area's produce market and increase the incomes of local farmers. Given that Connecticut had 22,000 farmers in 1944 and only 3,500 by the late 1970s, there was justifiable cause for concern that the end of the state's proud agricultural tradition was not far off. The hope was that the region's farmers could make more money selling at farmers' markets than they could as "price takers" at wholesale markets, while at the same time meeting the food needs of urban consumers.

For as long as anyone could remember, farmers had been taking their produce to the regional produce terminal market at the city's edge, where more recently they had been receiving less than the actual cost of production. The terminal market was the main receiving and distribution point for fresh produce moving in and out of the region, which increasingly over the past fifty years had been favoring the import side of the equation. Local farmers had continually lost ground and markets to national and global producers as locally produced food became passé.

In the distant past, viable retail public and outdoor markets had thrived in Hartford from the late 1700s through the 1930s. Ever since the lack of viable markets began driving the state's farmers out of business, Connecticut's Department of Agriculture had attempted to lure farmers back into the city. But farmers hesitated because they were wary of group-style direct-marketing concepts and were generally not comfortable selling at inner-city locations. To increase the likelihood of success, some early advocates had suggested that any new markets should be open only on the day that welfare checks were issued and that the markets be certified to accept food stamps. This idea was yet another cynical attempt to serve the needs of farmers first and those of city residents only as an afterthought. In 1978, things were different. This time, the farmers and urban residents were on equal footing.

FARMERS' MARKETS EXPERIENCE A GROWTH SPURT

Fast-forward thirty years, and the number of farmers' markets nationwide has swollen to more than 4,000. The USDA estimates that between 30,000 and 50,000 farmers currently sell at farmers' markets. They are one of the most important factors in the tidal wave of interest in local food and the mini-revival of small farms. Connecticut has more than 80 farmers' markets today. This growth accounts for the modest rise in the number of farmers in the state to 4,200, up 20 percent from the late 1970s. During that same period, the number of large U.S. farms has been steadily declining, so this growth is particularly notable.

Back in 1976, having ignored small- and mid-scale farming for many years in favor of large-scale commodity production (wheat, corn, soybeans), Congress passed the Farmer-to-Consumer Direct Marketing Act. This law provided state agencies and cooperative extension services with an incentive to develop a variety of new marketing approaches, including farmers' markets. West Virginia, Georgia, Louisiana, and Florida began expanding some of their state-supported wholesale markets to accommodate the growing interest in direct purchase by individual consumers. Pennsylvania's Department of Agriculture began serving as a go-between for farmers, food clubs, and consumer cooperatives, while also supporting the establishment of a dozen farmers' markets. California's Department of Food and Agriculture developed a certified farmers' market program after the state legislature enacted enabling legislation. The state launched more than thirty farmers' markets in 1976 alone.

The Greenmarkets of New York City, made famous by John McPhee's 1970s *New Yorker* story "Giving Good Weight," had the biggest impact on the explosion of farmers' markets. The Greenmarkets served as an inspiration and model for market aficionados everywhere, including those in Hartford. (With a nod to "Old Blue Eyes," one local advocate had said, "If you can run a good farmers' market in New York City, you can do it anywhere.") The name itself became synonymous with freshness, local farming, and vibrant outdoor markets. In fact, the Greenmarkets would become so protective of their name that they threatened a lawsuit against one New Orleans organization that simply wanted to call its new market a greenmarket rather than a farmers' market.

Despite the Greenmarkets' eventual success, the challenges that early organizers faced were enormous. They included finding acceptable sites, securing permission to use the sites (then retaining that permission for more than one year), recruiting farmers (some coming from as far away as two hundred miles), getting parking permits, towing cars, and dealing with the Gordian knot of New York City's bureaucracy. By 1981, however, New York City had sixteen farmers' markets and had spawned several others outside the city.

The Greenmarkets were a success because tens of thousands of New

Yorkers wanted fresh produce and dozens of regional farmers wanted new outlets for their crops. The success or failure of each market site was predicated on the quality of the location, in particular its proximity to densely populated neighborhoods and commercial corridors or nodes. Many markets were started, many failed, and many new ones emerged to take their place. But in all cases, the essential determinant of success was what worked best for the farmers—where they would make the most money and encounter the fewest hassles. In some cases, market sites were selected to meet the needs of a particular community and to serve mixed-income shoppers, but the expectation was that the more successful markets serving higher-income shoppers would support the less successful markets in lower-income areas. Ultimately, if the farmers made money, they would come. No money, no farmers. No farmers, no market.

Here again we find the conundrum that has faced market organizers since the 1970s. How do you make farmers' markets work for low-income shoppers—offering them affordable fresh produce at convenient locations—and also develop the markets as viable business opportunities for farmers? Although meeting both social and economic goals simultaneously is possible, it requires a deliberate effort, usually accompanied by a subsidy (either private or public), to be successful. Where the focus has been exclusively on securing the best return for farmers, markets rarely provide a good food-shopping opportunity for low-income people, because the food is priced too high and, more likely, the location is not accessible. When markets have been located exclusively to fit the needs of low-income communities, they rarely attract sufficient farmers to be successful unless a subsidy is available.

THE LOS ANGELES FOOD SYSTEM: UP FROM THE ASHES

The smoke-filled skies of South Central Los Angeles dominated TV news screens in 1992 as the impoverished community erupted in violence over the Rodney King case. The previous year, the beating of King, an African American, by white police officers in Los Angeles had been caught on tape. On April 29, 1992, a predominantly white jury acquitted the officers, sparking riots. The civil disturbance was yet another revelation that this racially polarized city had miles to go before social and economic equality would be achieved. In the wake of the violence and destruction, Los Angeles turned inward to assess its failures and create a road map to renewal.

Among the failures that it chose to scrutinize was the city's food system, particularly the segment that served Los Angeles's most economically disadvantaged neighborhoods.

A team of students from UCLA's Graduate School of Architecture and Urban Planning, led by faculty member Robert Gottlieb, was assembled to investigate what had gone wrong with the city's food system—in effect, to measure and assess the city's food gap. Their research uncovered not only a wide gap between the poor and the affluent but also a strong link between that gap and the civil disturbances. Their nearly 400-page report, "Seeds of Change," released in 1993, was one of most far-reaching and broadly constructed reports on an inner-city food system to date. It presented the following findings:

- Substantial levels of hunger and food insecurity existed in Los Angeles. Up to 27 percent of the residents in South Central Los Angeles did not have enough money to buy food.
- Cutbacks in federal, state, and local food programs had resulted in the emergency food system being overwhelmed. Twenty-five million pounds of food had been distributed in 1979; 450 million pounds in 1990.
- There was a high incidence of diet-related disease.
- Supermarket chains had abandoned the inner city as part of an industry-wide consolidation that affected all areas of Los Angeles, and some chains' reinvestment strategies for new city stores were inadequate and ineffective.
- Transportation to food outlets was a major challenge for many residents. Thirty-eight percent of the households surveyed did not own a car, and public transportation that served commuters did not serve food shoppers.
- The farmers' markets that existed in the Los Angeles metropolitan area provided high-quality, reasonably priced food to consumers and excellent returns to participating farmers, but they needed considerably more support to serve low-income communities effectively.

The report's recommendations called for a major reinvestment in the city's food system by both the public and private sectors, including a substantial expansion of supermarkets and farmers' markets in the city's most underserved areas. Indeed, one of the greatest frustrations expressed by South Central residents was the area's lack of high-quality, affordable food stores. Among the injustices that people shared was that the same high-quality, affordable food available in more affluent areas of Los Angeles was not available to them.

Progress toward implementing the recommendations, however, proved rocky. Supermarket chains made commitments to bring new stores into neighborhoods they had previously abandoned, but they often did not follow through on their promises. Nonprofit farmers' market organizers were sincerely committed to bringing the benefits of locally grown food (available in Southern California year-round) into the same neighborhoods, but in spite of their ability to establish successful markets elsewhere, their attempts in inner-city neighborhoods faltered as well.

In a follow-up report six years later, Andy Fisher, one of the original graduate students responsible for "Seeds of Change," focused on the lessons learned from attempts to organize farmers' markets in low-income areas of Southern California. His 1999 report, "Hot Peppers and Parking Lot Peaches," chronicled failed attempts in Van Nuys, Koreatown, Highland Park, and East Los Angeles. In all cases, organizers had conducted community outreach, received an impressive amount of media coverage, and located the markets in prime sites. But in the words of Marion Kalb, the head of the Southland Farmers' Market Association, "We aren't quite sure anymore how to organize farmers' markets in low-income communities."

NEW ORLEANS FARMERS' MARKETS SHOW THE WAY

The struggle for farmers' markets to help close the food gap would be played out almost every time a new market opened, an unsuccessful market closed, or a market underperformed. Even when markets were so successful that it seemed there was no room for improvement, a small voice in the back of the room would ask why there weren't more people of color shopping there, or why residents from the nearby low-income neighborhood weren't coming to the Saturday market. The answer came as market organizers learned the art of collaborating with multiple partners to achieve multiple objectives. In that vein, farmers' market leaders such as Richard McCarthy of New Orleans could sustain the value of a farmers' market without straining its capacity to deliver a wide range of benefits to a diverse number of people.

Spend a couple of hours with McCarthy, who works for Loyola University's Twomey Center for Peace Through Justice, and you soon sense that a stern Jesuit teacher and sophisticated gourmand have been fused not unpleasantly into one person. McCarthy, age forty-two, utters perfectly constructed sentences that are woven into equally fluent paragraphs and are spiced discreetly with the vocabulary of New Left politics. He has a well-formed theory of change that surgically dissects the failures of globalism and free market capitalism. He's scraped his knees on the rough ground of Old South politics, racism, and a city that is struggling to revive after the hellfire of Hurricane Katrina. He can hold forth with ease on the tenets of Marxism and the Mondragon Manifesto, but he prefers the simple pleasures of helping to restore people's pride in their place, heritage, and food.

As a New Orleans native, McCarthy has witnessed the South's decline, but more important, he has been an active participant in its renewal. Yes, he's been involved in political coalitions that have vigorously opposed the various escapades of David Duke and has watched Louisiana's partially successful attempt to crawl out of its petrochemical swamp and reverse its resulting high cancer rates. But rather than belabor these unfortunate legacies, McCarthy has adopted a fitting slogan from the early-twentieth-century Industrial Workers of the World labor movement as his guide: "Building a New World in the Shell of the Old."

In 1995, McCarthy led the way in the establishment of the Crescent City Farmers' Market, which would eventually become home to sixty Louisiana farmers grossing $2 million a year. Once he figured out how to organize the market correctly, he passed the wisdom on to other groups, which then started twelve additional markets in the region. One of the keys to his success was the creation of a mutually supportive system of rural producers and urban consumers. City consumers, including individuals, chefs, and a generous slice of the Big Easy's bon vivant class, began to see their fates intertwined with that of the farmers, who in turn, like farmers in Connecticut, were struggling to survive the changes in the food marketing and distribution systems.

McCarthy saw that New Orleans's survival depended in large part on the vitality of its food culture, so he organized the market as a creative economic engine that would channel the city's unique heritage in music, food, and conviviality. But along the way, he kept his feet on the ground so as not to leave the city's large low-income population behind. Job creation at all levels of the food chain and some food processing businesses, such as a neighborhood-based pasta company, were all part of the mix.

McCarthy was able to convince dairy farmers who were no longer making it in the low-priced commodity milk trade to turn their attention to an old-time Louisiana product called creole cream cheese, the taste of which is said to bring tears to the eyes of devotees. He built trust among farmers who had previously resisted coming to New Orleans without a gun because of the city's reputation as the murder capital of America. He shored up the confidence of a welfare mother who didn't believe she had what it takes to start and run her own business. And, most important for low-income consumers, he was able to pull the appropriate policy strings to release $100,000 annually in the WIC Farmers' Market Nutrition Program (FMNP) funds that provided a powerful incentive to thousands of low-income households to shop at the farmers' market.

McCarthy also knows how to connect people's hearts and minds to the market. When one of the farmers lost a leg in a farm accident, McCarthy spearheaded a drive among restaurateurs, shoppers, and the local media that raised $80,000 to cover the farmer's uninsured medical expenses. His broad vision of food and agriculture, coupled with his inclusion of the many groups he serves throughout the regional network of farmers' markets and the individuals affiliated with those groups, is the key to his success.

A MATURING PRESENCE AND A GROWING COMMITMENT

America's food consciousness was aroused from a deep slumber in the late 1960s. By the 1970s, hippies, health nuts, and overeducated young people, including myself, had developed an uneasy sensation that there was something wrong with our food system. This growing awareness incited a kind of disorganized guerrilla warfare that sent an ever-increasing number of disenchanted food fighters into food co-ops (sometimes called "food conspiracies"), organic gardens, farms, communes, and other virtuous attempts to avoid an existential crisis through food. While these early skirmishes probably inflicted more damage on the guerrillas than they did on their ill-defined enemy, they also produced a small class of skilled organizers and leaders capable of taking the struggle to a higher level. And in the battles to come, it was farmers' market organizing that established the first successful beachhead in the fight to overhaul the food system.

Steady growth in the number of farmers' markets has continued to the present. Although their impact is due more to their location than to their share of the U.S. food market—farmers' market annual sales are somewhat in excess of $1 billion and represent less than 1 percent of retail food dollars—they have established themselves as an extra-large presence in today's food consciousness. According to the Project for Public Spaces, almost two-thirds of all farmers' markets are located in downtown areas. More than one-third are sponsored by local government, which is a testament to their elevated standing as small economic and tourist attraction engines. While their commitment to serving all income groups is not universal, almost two-thirds of the markets participate in the FMNP, which provides special vouchers to low-income families to purchase fresh produce at farmers' markets (more on this later). And lest they forfeit their position as one of the country's food preference trendsetters, two-thirds of the markets have at least one certified organic farmer.

Farmers' market statistics tell only part of their story. People find something profoundly soothing about them, even grounding in contrast to the hurly-burly atmosphere of gargantuan shopping malls and the inhuman scale of big-box stores. Jon Carroll, in a column for the *San Francisco Chronicle*, captured this consumer sentiment best in describing a stroll through one of San Francisco's farmers' markets.

> I didn't notice any metal detectors, any armed National Guard soldiers, any signs warning about anything, no announcements over loudspeakers concerning unattended packages. And I thought that as long as there is really good produce around, it seems unlikely that people will think hostile thoughts. We need to sacrifice our best tomatoes to world peace, and go without ourselves. When a person sees his first good tomato after a long cold winter—well, that person is happy. That person is in a mood to compromise. A little tomato, a little cucumber, maybe some olive oil—come, let us reason together.

While consumers were increasingly finding the ambiance of farmers' markets both energizing and soothing, the farmers had barely learned how to display their tomatoes correctly before they found themselves carrying a lot of social weight on their shoulders. In the eyes of many, they had become the vanguard of the local food movement, veritable white knights in the battle against the industrial food system. They were dubbed, among things, the new standard-bearers for sustainability, the answer to the decline of downtown and Main Street, and the way to fill the gap left by supermarket abandonment of the inner city. This was obviously a lot to ask of any group of scruffy growers who parked their battered Ford 150s around the city square. As burdensome or even unfair as these responsibilities were, farmers' markets had become a source of inflated expectations, as well as powerful symbols of progress.

Some of the ambivalence toward farmers' markets and their role in promoting food security came into stark focus on September 11, 2001. Caught in the rain of destruction that followed the crumbling towers of the World Trade Center were the trucks of several farmers who were there selling their produce that day. (All of the farmers and their patrons escaped without harm.) As I watched the horror of those events on my television in Hartford—in a state that would lose more than two hundred of its own residents in the attack—I thought that if something as innocent as a farmers' market could be the victim of ideological warfare, there was no refuge from the globe's unresolved conflicts.

My depression deepened the following day when I learned of John Ogonowski, one of the pilots of American Airlines Flight 11, which struck the first tower. On his farm in Dracut, Massachusetts, Ogonowski had been mentoring Southeast Asian immigrants who had an interest in pursuing farming in the United States. He taught them how to farm in New England; he loaned them his land and equipment. One of nearly three thousand innocent victims that day, he had played a central role in developing the next wave of producers who might one day sell at farmers' markets. Perhaps he had the foresight to see that his own food security might be ensured if another generation of farmers was prepared to produce the region's food.

The events surrounding September 11 and their relation to the food gap were placed in perspective for me by Peter Mann of World Hunger Year, a national nonprofit organization based in New York City. He wrote of 9/11, "A hungry world is indeed a dangerous place. Only when our food policies begin with the hopes and dreams of the urban and rural poor will we build true food security, which will also be a huge step toward homeland security." Those words held both a challenge and an opportunity, implicitly recognized by local food advocates, that we all share the responsibility for freeing the world of hunger and food insecurity, and that our success in that task would make our own homes, workplaces, and communities more secure. In that respect, you might say that closing the food gap is just another form of enlightened self-interest.

Back in 1978 in Hartford, there was an explicit understanding that farmers' markets and local farmers could play a role in promoting the community's food security, even if it wasn't called that at the time. This first wave of "back to the city" direct marketers included an old Yankee grower from Simsbury, a retired machinist from Pratt & Whitney, a schoolteacher from Bloomfield, a part-time UPS driver from East Hartford, and several second-generation Italian fruit growers from Glastonbury. Selling good food, staying in business, keeping their farmsteads viable against the rising tide of suburban sprawl, and working together as families were their priorities, not feeding Hartford's growing number of poor people.

As the well-dressed employees of the city's downtown insurance companies besieged the farmers for sweet corn and peaches, some community activists asked if the market was serving its intended clientele. Some of the member organizations of the Hartford Food System, especially those that provided services to the city's low-income communities, expressed concern that the incomes, complexions, and residences of those shopping at this exciting new addition to downtown life did not reflect those of the people who were most in need. This criticism put the organizers and the Hartford Food System on the defensive. In a city that was growing increasingly sensitive to divisions between white and black, urban and suburban, and poor and affluent, the charge that the farmers' market was becoming just another entertainment venue for the privileged class was no small irritant.

In response to these concerns, the market's organizers secured the services of some faculty and students at Hartford's Trinity College to conduct a shopper survey to determine just who was using the market. The results indicated that the users were well mixed by income, race, and residency, and that 38 percent were either elderly people living on fixed incomes or low-income, non-elderly residents. The documentation quieted some critics, but the question of whom farmers' markets should serve was forever embedded in the Hartford Food System's agenda.

With the intent of bringing the benefits of local produce to every corner of Hartford, the organization launched a series of smaller neighborhood markets in 1980. The farmers would come from the same pool that sold at the downtown location. These markets would go through a series of ups and downs for years. A good site would be available one year but not the next; sales in a particular neighborhood were not enough to support the market; a community organization that promised to support the market with publicity and outreach failed to do so. By the mid-1980s, new farmers' markets began to blossom in the middle- and upper-class suburbs, and the farmers who had stuck it out in some pretty tough places were enticed by the prospect of making more money. There were many disappointments along the way, but the Hartford Food System persisted, and by the late 1980s, it had found the right mix of private and public support to realize the promise of farmers' markets for all residents of the city.

CHAPTER FOUR

Community Gardens

Growing Our Own

Here I have been these forty years of learning the language of these fields that I may better express myself.
Henry David Thoreau

I HAVE OFTEN FELT THAT both the good and bad experiences of growing up in New Jersey have influenced me over the years. In the face of so many things that were going wrong, the Garden State gave me the desire to pursue gardening, a respect for the livelihoods of farmers, and a passion for the protection of land and the environment. But in spite of a couple of aborted attempts to grow something in the backyard of my too-shady suburban home, it wasn't until I was twenty-three that I took a serious stab at gardening, in Massachusetts.

I'd like to think I've come a long way since then. The spring of 2007 marked my thirty-fourth consecutive year as a gardener. If memory serves me correctly, there have been twelve separate gardens in all, each with its own story and human entanglements, but always planted with the hope that I would finally get it right. With a microscopic eye on the plants and soil at my feet and a macroscopic view to the community around me, I have learned from gardens, with their kind of rough and honest grace, a host of necessary lessons. From my first garden, where I drove the crooked tines of Miss Wilson's spading fork into the sandy backyard of her tattered Wellesley Hills estate, to my most recent one in Santa Fe, New Mexico, where I guided a hefty rototiller through the heavy clay soil, I have approached each garden with the perennial faith that only gardeners know.

The author Michael Pollan says that a garden is "a place of many sacraments." The many hours of communion I have taken in the gardens of my past, hours spent on my knees in supplication to forces I barely comprehend and cannot control, testify to the truth of his statement. As I garden in community with others—which I have done off and on over the years, in urban landscapes as well as rural—these sacraments are no longer taken alone but are more often shared with others. Through the group action that is required to achieve something that benefits all as much as it does me, I have received unexpected gifts that are not available to the solo gardener.

CUTTING MY TEETH ON YOUTH GARDENING

Just out of college and having barely survived my own adolescence, I assumed the directorship of a municipally run youth program in Natick, Massachusetts, located about fifteen miles west of Boston. Not wasting time on a warm, Welcome Wagon kind of greeting, the town fathers advised me that if I wanted to have a job in six months, I'd better get "those damn kids off the street." Feeling highly motivated, I began to scout the region for "revolutionary" youth program ideas that, if replicated successfully in Natick, would assure me a long and prosperous career. Fortunately, I stumbled across exactly what I was looking for in the form of the Green Power youth farm in Weston, whose young people had found refuge from the mean streets of one of Massachusetts's most affluent communities by picking beans on town-owned farmland.

Green Power was based on a simple idea: put teenagers to work on a community farm, teach them the rigors of farm work, and instill in them the ethic of community service by distributing the farm's produce in Boston's low-income housing projects. Inspired by this model, but having no idea how one should farm (one local farmer had told me to just "dig it and dung it" but had neglected to tell me in which order), I set about the task of replicating Green Power in Natick.

Following two months of hectic planning—the clock was ticking on my six-month probation—I and a motley assortment of goodhearted volunteers started the Natick Community Farm. It officially began one chilly day in mid-April 1975. Resorting to a press-gang form of youth outreach that would be highly frowned upon in these more sensitive times, we rounded up twenty or so teens from juvenile police officers, school disciplinarians, and priests. We spent the first few hours shoveling horse manure from a local stable into town trucks that would haul it to a donated two-acre parcel of land owned by the local Audubon sanctuary. Having even less youth supervisory experience than farming experience, I quickly lost control of the situation. In less than an hour, my band of would-be agrarians were using their shovels and pitchforks to hurl great waves of fresh manure at each other.

From these inauspicious beginnings, we accumulated enough agricultural acumen to make the Natick Community Farm a respectable youth farming enterprise. I had managed to find a farm manager who had just graduated from the University of Massachusetts with a degree in agriculture but had no practical farming experience. We worked in the field together that first year, he standing with a plant science textbook in his hand and I on my knees in the dirt trying to implement his instructions as he read them aloud. Our lack of experience and the demands of dealing with an endless stream of adolescent issues did not allow for a peaceful communion with nature. Our stress levels were further elevated by the promise we had made to the young people that they would be paid from the proceeds of the farm's produce sales. Although this represented an incentive to work hard and share in the success of the communal enterprise (Soviet-era state farms used much the same model), it placed an additional burden on me. If we didn't make money, the young people would leave. If the young people left, the program would fail, and I would be looking for a new job (or be "reassigned" to Siberia).

For once, our ignorance paid off. Knowing nothing about the correct use of agricultural chemicals, the Natick Community Farm became organic by default. And as luck would have it, the short distance to Boston gave us easy access to that city's small but growing organic market niche. If we could produce organic fruits and vegetables of respectable quality, we could sell them to Erewhon Natural Foods Store for a pretty penny. And by some miracle, we managed to cultivate a righteous patch of organic cantaloupes for which Erewhon was willing to pay top dollar—if only we could deliver them ripe and ready to eat. The farm manager's college textbook, however, did not tell us how to determine when cantaloupes are ripe. Each day we waited in a heightened state of anxiety trying every trick we'd heard of to assess their maturity. After one particularly long day of thumping, sniffing, and squeezing melons, I lay my weary head down at night only to be awakened later by my frantic wife, who was yelling, "Stop it! Stop it!" I asked what was wrong, and she said, "You're squeezing my ass and shouting, 'Are the melons ripe? Are the melons ripe?' "

It was in those stony New England fields sloping gently down to the banks of the Charles River that my personal communion with nature became communal. Gardening was no longer a one-person enterprise—just myself at one with the plants, soil, and sky—but a social interaction that engaged the hearts and minds of many. Collectively with my young charges, and maybe like Thoreau before us, we began to shed our hurly-burly suburban skins in exchange for something a little softer, a little slower, a little more in tune with the demands of the farm. Subconsciously, we began to model ourselves after the cranky old Yankee farmers who would stop by the farm to visit. Curious and kindhearted, they would share stories about our field that made me believe that their knowledge predated the Pilgrims. And, of course, they would complain about the weather. "April's too wet to plant," they'd assert, pressing their huge, leathery hands firmly on their hips. "It means May will be too dry," they'd predict, squaring their immense bellies off against me, thereby deflecting any attempt to question their certainty.

These old fellows had names such as Bud and Chuck and exuded an inviolable wisdom that seemed hewn from the trunks of the nearby oak trees. Before long, we started walking and talking like them. "The frost will come early this year," one sixteen-year-old crew leader told me, drawing on her entire two months of agricultural experience for her prediction. Another young worker, who had earlier that season screamed at the sight of his first worm, confidently forecast a better fall broccoli crop than last year, even though he hadn't been there the previous fall.

Our experience and knowledge may have been shallow, but we were discovering the unique gifts that gardening and small-scale farming can bestow. Like Thoreau, we were "learning the language of these fields" so that we might better know ourselves, forge meaningful bonds with one another, and express ourselves more fully in our present community and beyond. Over time, our knowledge became rooted in that place, and our conversation became its vernacular.

The Natick Community Farm grew strong, self-reliant kids better than it grew food. Bud and Chuck would look at some of our bedraggled produce, shake their heads, and tell us that "farming is hard work," which was their way of demonstrating unqualified compassion. As successful as the farm was as a youth project, it did not succeed in removing "menacing gangs" of youths from the nearby malls (although one local policeman did credit us with a marked reduction in hubcap theft). But the town fathers were so smitten with it as a focal point for wholesome and productive activity that they were convinced we had single-handedly wiped out illicit drug use.

Over our first five years, we expanded the scope of the Natick Community Farm to utilize fully the twenty-plus acres of land that the town had granted the farm. We built a maple-sugaring house out of wood recycled from a demolished home. The young people learned to make syrup from the sap we collected from the town's sugar maple trees. Ushering in the age of alternative energy, we constructed a solar greenhouse to grow vegetable seedlings. We saved a beautiful barn, built in 1840 from hand-hewn chestnut beams and located on the farm's property, from demolition. We turned the restoration into a major youth project that was funded by Natick's 1976 bicentennial committee. To return the barn to its original purpose and give the young people a greater range of farming activities, we developed pastures and introduced a Noah's ark of farm animals.

Still operating to this day, the Natick Community Farm has since constructed new greenhouses, living quarters for staff, and classrooms for educational activities. The program has stayed true to its original organic principle of "dig it and dung it" by becoming a regional center for teaching and advancing sustainable agriculture. But how do you measure the impact of a farm like this? Each year it produces a few thousand dollars' worth of produce, fifty to seventy-five gallons of pure maple syrup, a few hundred dozen free-range eggs, and a few hundred pounds of meat from some combination of pigs, chickens, and lambs. Some food no doubt has found its way to the local food pantry. The farm's output has been about what one would expect from a small New England farmstead, hundreds of which have gone out of business every year.

The value to the farm's young people has not been entirely what the town fathers had in mind. The farm has motivated very few of its participants to pursue careers in farming. For many unfortunate youths who were emotionally crippled by failed families, mental illness, or chronic drug use, the farm has not provided a therapeutic environment powerful enough for them to turn their lives around. It has, however, become increasingly important that the town's young people have an alternative frame of reference that doesn't include the local mall and that gives them a respite from an economic system that treats them as if they are only consumers-in-training. A hand plunged into freshly tilled soil, the squiggle of an earthworm, or a tiny seed mysteriously transformed into a green life force were, at the time of the farm's inception, experiences fast approaching the brink of extinction. The Natick Community Farm and similar projects have amply demonstrated that life offers a richer menu of choices.

BUT IS GARDENING FOR EVERYONE?

Being the hardy and self-reliant people we think we are, we harbor a belief that when all else fails, we can always return to the soil for our sustenance. Should the oil run out, the trucks and trains stop running, and the far-flung food network crumble into dust, the land and our instinctive ability to coax food from it will save us. At least this is the folklore we fall back on, a wisdom passed from our pioneering ancestors, whose skill, knowledge, and work ethic we would have to channel in order to truly live off the land again. Having witnessed many sincere but ultimately failed attempts to transform dirt, water, and seed into food, I tend to look somewhat askance at those who suggest that more of us, if not all of us, and especially the poor, should "grow their own." Whether we invoke the agrarian spirit of Thomas Jefferson, the sod-busting pioneers of the Iowa plains, or the Victory Gardens of World War II, our claims of self-reliance often come precariously close to self-righteous pontificating.

Catherine Lerza's Hartford report (see chapter one) made it clear that promoting community gardening in particular and urban agriculture in general were extensions of time-honored traditions that date back to the Middle Ages. "In Europe," Lerza argued, "the custom has never died; garden plots are at a premium in cities in Britain, France, and Scandinavia. During World War Two . . . Americans were encouraged by the federal government to be as self-reliant as possible; victory gardens sprang up in schools, churches, parks and private homes." But once the culture of scarcity was transformed to one of abundance, which was the case following the war, "urban gardening declined sharply."

As far back as the 1890s, there are examples of communities coming together to form gardening associations to promote the use of vacant lots and other open space to ameliorate the hardships of economic downturns. Perhaps the forefather of gardening as a means of economic uplift for the poor was Detroit mayor Hazen Pingree, who during that period suggested gardens could be a form of charity with a better return on investment. But revealing something of the true sentiment behind this effort, Pingree touted gardens as a way to cast off the moral malaise that he and others no doubt felt inflicted the poor. He said that one of the benefits of gardening was "that the needy are therefore assisted without creating the demoralization in the habits of the people that gratuitous aid always entails." In a community gardening project called "Potato Patch Pingree," 40,000 bushels of potatoes were harvested in one season. Even though the whole idea of urban gardening as a form of hunger relief was derided by editorial writers of the time, nineteen cities, including New York and Philadelphia, had similar gardening projects by 1898.

In addition to supplying low-income residents with healthier and more nutritious food, community gardens have been tagged with many other beneficial characteristics by their advocates. Among those cited by Jerome Kaufman and Martin Bailkey in their paper "Farming Inside Cities" are reducing the amount of vacant and unproductive urban land, improving the public image of troubled neighborhoods, increasing the amount of neighborhood green space, developing pride and self-sufficiency among inner-city residents who grow their own food, and providing jobs for youths and adults.

The assumptions about vacant land and job opportunity benefits are interesting. They suggest that an activity such as community gardening is designed to assist poor people or the neighborhood where they reside and that it is based on the need to put something that is of no value to anyone else to use. Indeed, the amount of vacant land in many of America's cities is, according to Kaufman and Bailkey, astounding and a painful testament to the decline of urban areas in this country. In 2000, Philadelphia had 30,900 vacant lots, an increase of almost 100 percent since 1992. New Orleans had 14,000 vacant lots, and at least one-quarter of the properties in most of Chicago's poorest areas were abandoned. With the loss of half a million people since World War II, St. Louis has assumed control of 13,000 tax-delinquent parcels totaling 1,200 acres. Americans abhor waste, whether it's land, food, or simply open space, and what better way to use something that has fallen outside the standard utilitarian economic model than to feed or employ the poor?

At the time the Hartford Food System was getting under way in the late 1970s, Lerza reported that there were 20,000 community gardens in the United States operated by city, state, and federal government and by private nonprofit organizations. Kaufman and Bailkey reported 6,000 community gardens in the late 1990s. What accounts for the discrepancy is not known, other than that I have never met a local official who has an accurate sense of how his or her city's vacant land is being used.

Starting in the 1970s, community gardening in Hartford was supported by the Knox Parks Foundation, which secured sites, made minimal physical improvements (such as bringing in an initial load of topsoil and compost), and organized the allocation of plots. Including Knox in the Hartford Food System was an attempt to give all its members access to Knox's technical expertise, while also putting more resources and effort into expanding gardens into underserved low-income communities. Lerza projected that gardens would save their participants money on food, while also providing seasonal jobs for youths, whose wages would be paid by city and nonprofit agencies (which usually received funding for summer youth employment programs from the federal government). The hope was that community gardening in Hartford would join the expanding national gardening movement and begin to close the food gap, improve residents' quality of life, and create educational and employment opportunities for unemployed or underemployed people, primarily youths.

The hope exemplified by Hartford was not unlike the quasi-utopian vision that often propelled community gardening enthusiasts and their more ambitious cousins, urban farmers. Their language and dreams often suggested that rubble-strewn lots could be turned into oases where the urban desert could bloom. Although something approximate to this has happened on occasion over the years, the reality has generally been less paradisiacal. Granted, a little patch of green sprouting in an otherwise unforgiving urban landscape is desirable for many reasons, not the least of which is the relief it gives the eye. But as Jack Hale of the Knox Parks Foundation readily admits, Hartford's community gardens have made only a marginal contribution to the city's food security, with the exception of a relatively small number of ardent gardeners who have significantly augmented their food supplies.

That being said, it has proven worthwhile for communities to make a public commitment to providing land, horticultural training, soil and compost, and other means of support to enable people who want to garden to do so. Whether people are motivated by the myth of self-reliance, the fear of a cataclysmic event, or simply the wish to make something ugly into something beautiful, society should permit them to stand in humble repose on their own tiny plots of land and to make what magic they can of it. Doing so affords them the opportunity to come together in community to grow plants and to experience for themselves the pulse of the seasons marked by the productions of the earth.

GARDENING IN HARTFORD

Energized by my experiences in Massachusetts, I came to Hartford with the intention of bestowing the benefits of gardening and healthy food on a place that suffered from a paucity of both. The first of many Hartford gardens that I would help develop was in the city's Bellevue Square neighborhood in 1979. Unlike middle-class Natick, Bellevue Square had some of the harshest demographics in the Northeast—desperately poor and crime-ridden, and a poster child for every failed urban redevelopment scheme in America. Its only visible asset was an abundance of vacant land, which was no more than a sorry testament to failed attempts to build something on it. The fact that the residents expressed a desire for a community garden suggested how universal the human craving for a little patch of dirt is. But in this neighborhood, the patch of dirt was a rubble-filled lot, and if you stuck your hands into the soil, you might get stabbed by a hypodermic needle.

Once residents chose the site for the garden, I worked with them to arrange for water, fencing, and rototilling. To bolster the hardscrabble medium that would grow their plants, we imported hundreds of cubic yards of leaf compost generated by the shade trees of Hartford's affluent suburbs. On the day the big trucks arrived to dump the black compost on our site, I stood with the residents, the only white face in a sea of black ones, determined to demonstrate that I was prepared to toil with them to reclaim this godforsaken piece of ground.

The group included a neighborhood youth organizer who had achieved legendary status by risking his own life to separate two warring youth gangs. Armed with nothing but words and the respect he had earned from many years of living in that community, he convinced them to lay down their guns and submit to a negotiated settlement. The gardening group included many elderly African American men and women as well, most of whom had roots in the rural South. While they couldn't do too much heavy lifting, they provided many lifetimes of gardening advice to the younger people who could. One stout elderly woman, who stood as tall as I in spite of her severely stooped shoulders, comes to mind. If my memory serves me correctly, she would later be the first person to give me a bunch of collard greens to take home, but only after I listened carefully to her detailed cooking instructions.

And then there were the children, many and rambunctious, who came in all ages, shapes, and sizes. Most of them wouldn't last too long, throwing a few half shovelfuls of compost before losing interest and running up and down the compost piles until they were scolded by their mothers. But one little fellow stood out for his stamina and determination. Standing barely as high as my waist and skinny as a rake handle, he pushed and shoved that compost around with the ferocity of a pit bull. It might have been my imagination, but something told me that he was doing this to impress his father, a large, hard-looking man who neighbors told me had done time at the state pen.

I enjoyed being there, but after a while I felt like a guest who had stayed too long. I soon realized that I was just one more well-intentioned white guy who had to learn that Bellevue Square was not his place. It belonged to the people who lived there and who used their hodgepodge assortment of tools that day to make it a little better. I was welcome to visit, to provide resources, even to bring newspaper reporters by to write clichéd stories of how this poor community was planting the seeds of its own revival. But the residents of Bellevue Square, like the seasons of the year, had their own rhythm. As the sons and daughters of southern sharecroppers, or Puerto Rican families who had migrated to the Connecticut River valley to pick tobacco, they had a richer store of agricultural knowledge than I did. The garden flourished and eventually won the city's best garden of the year award. The gardeners taught me how to grow and prepare collard greens and cilantro, and I gained satisfaction from watching them succeed on their own terms.

A COMMUNITY GARDEN OF MY OWN

On the banks of the Park River, which runs along Hartford's western border, on a rich piece of bottomland called the Watkinson Community Garden, I finally staked a claim to my own 20-by-20-foot community garden plot in 1996. It was here, among my neighboring gardeners, that I coaxed Brandywine, Rutgers, and Roma tomatoes from the black, stoneless earth. Sunflower heads as wide as Frisbees reached eight feet high and shaded me as I weeded peppers, onions, and dahlias. The vines of Romano bean plants wound their way up the jury-rigged tepee I had constructed from saplings cut from the riverbank.

The Watkinson Community Garden was unique among the fifteen or so Knox Parks Foundation gardens. It was much larger than any of the others, containing well over one hundred separate plots. The land was controlled by a nearby private school, which seemed happy to let us use it for gardening. More important, it sat on land that could never be developed because it was part of a floodplain. Indeed, the Park River would occasionally overflow its banks in the spring and send a rush of water over half of the garden plots, leaving them covered with a foot or more of water until late April. This annoyed the more avid gardeners, who were ready to put their peas in the ground come hell or high water.

Besides the deep deposits of alluvial soil that the Park River had so generously laid down year after year, the Watkinson site and its adjoining areas of meadow and scruffy woodland was one of Hartford's best-kept secrets. Cardinals, finches, orioles, bluebirds, swallows, and killdeer seemed to burst from every direction, creating a constant blur of color. Deer, which leaped our meager fence without hesitation, took out a row or two of lettuce seedlings before moving on. Muskrats burrowed into the riverbanks. The unfortunate woodchuck that made the mistake of staying too long inside the perimeter and the wayward black snake that consumed many unwanted rodents all made this little corner of Hartford the most vibrant urban wildlife habitat I've ever encountered.

But the most interesting features of the Watkinson Community Garden were the gardeners themselves. This site tended to attract the city's more serious gardeners. In Hartford, that meant middle-aged Jamaican men (the city has one of the largest African Caribbean communities in the country), for whom gardening was both an economic necessity and a social pastime. Raised in the more formal British style predominant in Jamaica during their youth, they addressed each other as Mr. Marley or Mr. Bennett or Mr. Nelson. I never learned their first names because this was how I came to address them as well, a sign of respect. They even addressed me as "Mr. Win-nee," which, when said with a Jamaican accent, had a decidedly silly ring to it.

Their beer of choice was Heineken, which they started offering me after they determined that I had a modicum of gardening skill. We would stand around drinking cold beer on a hot day while trading gardening tips, seeds, and seedlings, as well as produce when the harvest threatened to bury us all. I welcomed their "peas" (dried beans that they would store for the winter) and hot peppers, but I never developed a taste for callaloo, a spinachlike leafy green that Jamaicans regard with reverence. I nevertheless admired the sight of these men proudly bearing hefty four-foot-long, 18-inch-diameter bunches of just-cut callaloo over their shoulders, the stalks bound tightly with twine. They would load several bunches at a time into the trunks of their old cars and sell this much-sought-after vegetable to the Caribbean restaurants in the area, as well as to their neighbors. I was able to return their generosity by sharing my tomatoes and onions. They would rub the dirt off an onion with their hands and eat it on the spot, just like an apple. But they would accept my gifts of broccoli, a vegetable they had not cultivated a taste for, only if they thought that they would hurt my feelings by not accepting it.

While these courtesies and customs were generally the norm, my Jamaican friends would often explode into a raucous, almost violent kind of teasing and screaming, all delivered in a patois that was wholly incomprehensible to me. On more than one occasion I bolted upright from my weeding, expecting to see someone being beaten with a garden rake. To my chagrin, I would then realize it was only their way of showing good fellowship toward one another.

There was something about their intensity that drew me to them—an intensity that, at least when it came to gardening, matched my own. Their most commonly used garden tool was a machete, a rather fearsome-looking implement the likes of which I've never found in a Smith & Hawken catalog. They would prepare their plots, plant their seeds and seedlings, weed, and harvest with nothing more than a machete. Every stroke was applied with maximum force and surgical precision, and they were able to keep at it for hours on end.

Early one June, Mr. Bennett discovered that a rather large woodchuck was encamped in the middle of the community garden and feasting handsomely on the new spring greens. The men mobilized immediately with the precision of a well-trained military unit. Three or four volunteers went on reconnaissance to locate each of the woodchuck's access and egress points and to stand watch with hoes and shovels at the ready. One or two "sappers," armed with a jerry can of gasoline, found the main hole and filled it with gas. Yelling "Fire in the hole!" one man dropped an entire book of lit matches into the poor woodchuck's den, which immediately exploded in a ball of fire that rose ten feet into the air, throwing the gardener turned warrior hard onto his back. In no time, the singed but still agile 'chuck darted from one of his exits, only to be greeted by a shovel-wielding Jamaican, who was soon joined by his comrades. Flailing their arms and gardening tools—plowshares turned momentarily into weapons—and egged on by a cacophony of inflammatory patois, the men soon brought the furry fellow to an unsavory end.

While they would never speak in patois to me or invite me to join their wild, table-pounding games of dominoes, I did manage to achieve a kind of rough acceptance when the men felt confident enough to criticize my gardening methods. On my knees in the dirt one day, digging in my tomato plants, I looked up to see Mr. Marley and Mr. Bennett looking down at me, shaking their heads disapprovingly. I asked if there was something wrong. They said, "Mon, ya don't plont tah-mah-tows that way!" and then proceeded to demonstrate what they considered to be the correct technique. I was in a quandary. Should I continue to use the tried-and-true method I learned as a child from the Swiss-German Ilsa Wirth and refined in my late twenties in Natick under the tutelage of Bud, the seventy-year-old Yankee farmer? Or should I take the advice of these two Jamaican men now regarding me sternly? Suddenly, I remembered what my colleague Jack Hale had said: "The most important word in *community garden* is not *garden*." I now plant tomatoes the Jamaican way.

GARDENING IN THE AFTERMATH OF A HURRICANE

The power of community gardening and other similarly organized small-scale farming efforts in nontraditional areas such as urban America is not found so much in the rate of return to the food supply but in the rate of return to society. And while the contribution of these alternative enterprises to food security might be marginal (my Jamaican friends and others with strong agrarian roots and well-honed gardening skills notwithstanding), their ability to build and even rebuild communities deserves note.

Nowhere was this more evident to me than in the Ninth Ward of New Orleans, where I worked as part of a volunteer team to restore community gardens in March 2006, only six months after Hurricane Katrina. The gardens, like the neighborhoods, homes, and churches around them, had been underwater the entire month of September 2005. The accounts of the hurricane's aftermath—the complete and utter abandonment of large neighborhoods, the giant trash piles, and the parking lots full of wrecked cars—were heartbreakingly accurate. Never have I seen a place, formerly home to tens of thousands of people, look so forlorn. Perhaps the most striking sight was a solid black stripe that ran along the side of almost every wooden house, running perfectly straight from house to house, block to block, six feet above the ground. This line was Katrina's high-water mark, a defiant symbol of nature's power.

Six months after this catastrophe in a neighborhood where nearly every house and building was still abandoned, a few of us outsiders had gathered with some of the community's faithful to put our hands together in the dirt. Behind a church that was providing a variety of relief services to the homeless, we amended the ground in an area measuring less than a quarter of an acre with imported soil and manure in an attempt to restore the fertility Katrina had drained away. We weeded, hoed, raked, and planted tomatoes, peppers, eggplants, and even citrus trees. Our actions had a certain determined, in-your-face quality to them, as if to tell Mother Nature, "I saw your black line. Well, here's our line, and we dare you to cross it!"

Later, we moved on with Macon Fry—a lanky, angular man in his late forties, one of the garden leaders from the New Orleans Food & Farm Network—to another community garden a mile away. Macon and some neighbors had restored this site, located in a neighborhood no more repopulated than the last one, the week before. The carefully combed black earth had been meticulously shaped into raised beds that were already sprouting green sprigs of lettuce, Swiss chard, beans, and cabbage. It was certainly a blessed little spot, one that beckoned to all who passed, like a small wedge of paradise surrounded by the inferno that Katrina had wrought. At the garden's edge sat an old wooden garage that had been shoved by Katrina's wind and water and now was pitched at a 60-degree angle directly over several rows of Swiss chard seedlings. If anyone had leaned against the structure, it would have crashed into the garden. When I asked Macon why they had planted their vegetables in such a risky place, he said, "We had to start somewhere to feel like we belonged here again. I guess you could say it was an act of faith."

GARDENING IN THE AFTERMATH OF A RIOT

Hurricanes, toxic soils, vandalism, public indifference, and landowners who would rather turn undeveloped sites into parking lots are just some of the obstacles that community gardeners face. It takes extraordinary resolve on the part of community members to develop new garden sites, but it also takes hard work and guts to hold on to them. Leadership is one determinant of success for community gardens, just as it is for farmers' markets, food co-ops, and a host of other community-based food initiatives. When a single person emerges who knows the neighborhood, knows how to make city hall listen, and has the respect of the community, success will most likely follow.

One such person was Helen Johnson, an extraordinary African American woman whom I had the opportunity to meet at her South Central Los Angeles community garden in 2003. My first impression of her was that of a favorite grandmother—smiling, caring, and, at age seventy-two, sufficiently spry to wear out a gaggle of active grandchildren. Sitting with her in the community garden she organized, I was alternately comforted by Southern California's balmy breeze and Johnson's loving warmth. Though tender on the outside, she had an inner core as tough as the gritty landscape that surrounded her small patch of paradise.

The Vermont Square Community Garden straddles Vermont Avenue near Forty-seventh Street, one of the busiest traffic corridors in Southern California. Not far from the spot where collards, carrots, and begonias now grow, the South Central riots rocked Los Angeles and shocked the nation in 1992. Like similar events that have ripped apart communities elsewhere, this explosion was ignited by the bad chemistry of poverty, injustice, and distrust. Catastrophes such as these often plunge residents into a black hole of despair, leaving them bereft of any hope that the scales of justice might one day be balanced. But for leaders such as Helen Johnson, the ensuing chaos gives them the gumption they need to undertake a project that will make their neighborhoods better places and inspire their neighbors with a vision of a safe and healthy community.

Johnson liked to tick off her opponents, especially when it came to holding city hall accountable. Her first community action occurred in 1988, when she persuaded the city to install locked gates at the head of alleyways that had become dumping grounds for all manner of debris. After the riots, Mayor Richard Riordan asked her to be an area chairperson. This led to efforts to rid the neighborhood of liquor stores, which had sprung up like the weeds in the sidewalk's cracks. And then there was the community garden.

From an abandoned right of way that had formerly been used by LA's trolley cars, Johnson and her neighbors carved out two plots on either side of Vermont Avenue. With funds from an area foundation, technical assistance from the Trust for Public Land, and volunteer lawyers who were members of the nearby Temple Israel, her newly formed nonprofit group bought the land for $150,000 from the Metropolitan Transportation Authority (MTA). Johnson proudly proclaimed that theirs was the "first garden to be owned by a Los Angeles community."

A stroll through the garden today takes you down paths lined with fruit trees, flowers, and vegetables growing in raised beds. There are compost piles in various states of decay, mixed with horse manure from the nearby Hollywood Park racetrack. A gazebo, built with funds that Johnson talked out of the MTA and with lumber donated by Home Depot, dominates the garden's front face. And a picnic table that has no doubt been party to a thousand hours of delightful chatter commands the garden's center stage.

More than thirty gardeners, all of them over fifty-five years of age, tend their gardens with loving care. The grandparents bring their grandchildren, who learn the joys of gardening and the pleasures of eating vegetables. Teenagers from the Green and Clean youth program come by regularly to help the seniors tend to needed improvements and repairs. The site is beautiful, almost idyllic, and stands as another testament to the universal desire to wrest something peaceful and lovely from an environment that was hostile and blighted. The experience of developing the garden, purchasing it, and tending it had, in Johnson's words, "brought the community closer together."

According to Johnson, the food from the garden was "a supplement and necessity for those who live off those once-a-month checks." She had seen the face of hunger in the schoolchildren she looked after as a childcare worker in the local schools. In addition to her vision of a better community, it was that image that kept her going 24/7, engaged in a never-ending string of volunteer activities, including the emergency food pantry she ran out of her church as well as her home. "I know some of those who ask for food, and others I don't know, but all I need to know is that they are hungry," she said.

This corner of the world lost Helen Johnson when she passed away in 2006. She was a model community leader, putting others before herself and encouraging people to help themselves. She kept the big picture of a better community in mind, while also attending to the thousand details that keep any community going. And she always challenged the system. Her mantra, you might say, was, "If you see something that's wrong and keep quiet, then you're just as guilty as the person who's doing the wrong."

Johnson left her mark on almost every block of this community. The Vermont Square Community Garden, like many other good things in this tough corner of the world, would never have existed, let alone thrived, without her.

A LOOK BACK AT LESSONS LEARNED

Napoleon apparently told his personal secretary to delay requests from his subordinates for at least three weeks. His reasoning was that people usually figure out how to solve a problem or resolve an issue on their own. Jack Hale said that part of his job as a community garden organizer was to drag his feet for as long as possible when gardeners made requests. "That gives them time and motivation to figure out how to solve the problem themselves. The more gardeners become dependent on us, the weaker the community garden will be."

This might sound like tough love, or even bureaucratic incompetence, but as people struggle together, they inevitably forge bonds that are difficult to break. I once heard Marty Johnson (no relation to Helen Johnson), the executive director of a Trenton, New Jersey, community development organization called Isles, Inc., say, "When you help somebody who is fully capable of helping himself, then you aren't helping them." Johnson knows what he's talking about. Isles is one of the most effective community development organizations I have run across, and it also has done a great job of organizing community gardens. Anyone who has spent countless hours organizing one of the many failed community gardens across America would do well to listen to Johnson. As one of those who didn't heed that advice as often as I should have, I have come to believe that community gardens can help people fill the food gap only when they are motivated and encouraged to do the hard work that forms the building blocks of community.

Having said that, anyone who has worked in an urban environment in some form of gardening or agriculture is aware of the extraordinary challenges that city farmers face. Jerome Kaufman and Martin Bailkey identified some of these in "Farming Inside Cities."

- There is a great deal of skepticism toward urban garden enthusiasts ("How can you possibly expect to grow healthy food in the city?") and urban farming in general. In most people's minds, food production is associated with rural areas, not vacant city lots.
- There is a lack of funding for urban gardening enterprises, especially to cover start-up expenses associated with site improvements, which can sometimes be quite high depending on the site.
- Urban gardening is rarely seen as the best use of vacant inner-city land by government officials, whose first choice for land use is residential or commercial development. One of the biggest difficulties that the Hartford Food System and Knox Parks Foundation faced was securing permanent control of, or even a long-term lease for, a community garden site. Whether the landowner is a public or private entity, it is rarely inclined to tie up land for a use that will generate little or no income.
- Toxic soils, or the fear of such, make people uneasy about using urban land for food production. Site testing is almost always advisable for any new garden site, but there are also mitigation methods that can make any land short of an EPA Superfund site safe for gardening.
- Crime, vandalism, and petty theft can be major obstacles. There is nothing more heartbreaking than an earnest, hardworking gardener who arrives at his plot one evening only to find all of his beautiful vine-ripe tomatoes stolen.
- Some cities, especially during much of the 1980s and 1990s, have been hard-pressed to provide even basic services such as garbage pickup and police protection. Community gardening is regarded by some people as a frivolous endeavor in light of more serious and pressing demands.
- Gardening skills are not acquired overnight, and many first-time gardeners are discouraged when their plants and crops don't look like those portrayed in the seed catalogs. A little technical assistance is often necessary to give the neophyte gardener the resolve to try gardening for at least two seasons.

I can attest to the pleasure and pain that are the opposite sides of the gardening coin. For community gardeners to be successful in their rugged urban environments—to say nothing of making more than a minor contribution to closing the food gap—cities must make a serious commitment to providing land that is suitable for gardening. Most important, that land should be available for at least five years. Adequate funding, from public or private sources, must be available to defray some of the start-up and infrastructure costs (fencing, plumbing, and topsoil). Training and technical assistance are essential not only to help gardeners overcome emotional setbacks such as bug-infested plants and poor-quality crops but also to provide an appropriate amount of organizing assistance so that *community* remains the most important word in *community garden.* When done right, community gardening is one of the most satisfying endeavors in life.

CHAPTER FIVE

Food Banks

Waste Not, Want Not

Whoever degrades another degrades me;
And whatever is done or said returns at last to me.
Walt Whitman

LIKE MANY CHILDREN OF MY GENERATION, I was reminded of the perilous state of the world by my mother's frequent admonitions to "clean my plate" or she would send my meals to the "starving Europeans." My subsequent review of 1950s European history suggests that there were indeed Europeans who were hungry, and perhaps there were even some remnants of World War II–induced food insecurity. But I suspect that my mother was less motivated by a charitable impulse to share her son's meager rations with the rest of the world than by her own experience with scarcity during the Great Depression. As has been pointed out already, food was often scarce, unevenly distributed, or not affordable during the 1930s, which placed many Americans in a state of food insecurity and even hunger. And when scarcity is a reality, people will eat every scrap of food that's available, even if they don't like it or aren't physically hungry at that moment. In other words, the belief that scarcity is lurking just around the corner makes the thought of wasted food abhorrent to many Americans, especially those, like mother, who lived through periods of extreme belt-tightening or even worse.

Up until a couple of years ago, you could go to the homepage of America's Second Harvest's website and find a counting device that told you up to the minute how much food had been wasted in the United States in that particular year. The numbers increased at the rate of more than 3,000 pounds per second, and, for example, as of November 16, 2004, at approximately 1:30 p.m. Mountain Standard Time, 84,309,267,423 pounds of food had been wasted since January 1, 2004. Immediately below the number, you could click on Stop the Waste, which appeared in bold red letters and linked you directly to the Second Harvest online donation page. What intrigued me most about this link was that the appeal to the potential donor was not about hunger—the number of hungry or food insecure people in the United States or the number of people going to emergency food sites—but rather how much food we wasted. In other words, the nation's largest antihunger organization believed that our sense of moral outrage was more likely to be heightened by our national profligacy toward food than by the existence of hunger in the world's wealthiest nation. And in all likelihood, America's Second Harvest's assessment of what motivates Americans in this regard was correct.

While the need for food in low-income communities has been a perpetual and driving force behind the spiraling growth in emergency food programs from the early 1980s right up to the present, the institution of food banking evolved as one of codependency between food donor and food recipient. The argument goes like this: the food bank needs food to give to its clients, and food donors—food manufacturers, restaurants, retailers, individuals, farmers—must dispose of food they can't use or sell.

A statement from Second Harvest's updated website has this to say about the role of food donors: "The distribution [of two billion pounds of food every year] is made possible by companies in the food and grocery industry who regularly donate surplus, distressed and unsaleable food and grocery items to those Americans who need it most." The condition and quality of this food varies so greatly that many volunteers (up to two-thirds of all food-banking personnel hours are provided by volunteers) are needed to sort, clean, bag, and distribute the food. In the course of doing this work, everybody from the volunteers to the professional staff to the donors themselves becomes heavily invested in the organization's twin mission: feeding the hungry and preventing the waste of food.

The risk in all of this, of course, is that the multibillion-dollar system of food banking has become such a dominant force in the antihunger world, and so tied to its donors and its volunteers, that it cannot step back and ask if this is the best way to end hunger, food insecurity, and poverty. Part of the problem is that this highly effective mobilization of donors, volunteers, and other community resources has skewed their view of the recipients. Researchers in Toronto who studied the relationship between volunteers, the local food bank, and the recipients uncovered some disturbing attitudes. As one volunteer said indignantly, "We're not a grocery store . . . I think [the clients] should be satisfied with whatever they get." The question arises, who is serving whom? Are the food banks serving the food donors? Are the food banks serving the volunteers? And how well are the poor being served?

All of this apparent ambiguity makes perfect sense to me. In 1982, as I was wracking my brain to stem the growing tide of hunger in Hartford, several organizations and faith communities started the Greater Hartford Foodshare Commission, which would later become the third-largest food bank in New England. The food bank increased the city's capacity to receive and distribute donated food. And as the need grew and the number of sites where emergency food was available increased, it became increasingly adept at securing more food—necessity being the mother of invention, as they say.

No donation was too small, too weird, too disgusting, or too nutritionally unsound to be refused. I remember the load of nearly rotten potatoes that we gratefully accepted at the warehouse's loading dock and then promptly shoveled into the dumpster once the donor was safely out of sight. One of our early board meetings included a cooking demonstration by a group of local entrepreneurs who were trying to develop a market for horse meat. The product's name was Cheva-lean, which of course was taken from the French word for horse, *cheval*. The promoters reminded us that the French, the world's leading authorities on food, ate horse meat and, therefore, that our poor clients could certainly do the same. And to top that, I still have the recipes from the University of Maine Cooperative Extension Service that helped us use the moose parts that were proudly donated by representatives of the Connecticut Fish and Game Division when an unfortunate moose met its end jaywalking across I-84.

We did our job well, and everything grew: the warehouse expanded, the volunteers multiplied, the food and monetary donations poured in, and the demand soared. Foodshare bent over so far backward to accommodate food donors that at one point, long after I had left, its mission statement changed from a simple affirmation of its desire to end hunger to one that emphasized the need to manage food waste. (That seemed to be the prevailing public relations wisdom in the food bank community at the time.) I remember thinking how odd it was that a highly reputable charitable institution should exist for the purpose of taking care of comestibles that nobody wanted.

Much later, I was grateful to see that the food bank dropped this unseemly reference to waste and refocused its attention on the hungry. More important, it began collaborating with antihunger advocates to seek longer-term, systemic solutions to Hartford's food insecurity problem. Food banks, however, remain a dominant institution in this country and assert their power at the local and state levels by commanding the attention of people of goodwill who wish to address hunger. Their ability to attract volunteers, raise money, and conduct capital campaigns for never-ending expansions of their facilities approaches that of major hospitals and universities. While none of these actions are inherently wrong, they do serve to distract the sensitive public and sympathetic policymakers from the task of harnessing the necessary political will to end hunger in the United States. And as Janet Poppendieck makes clear in her book *Sweet Charity?* there is something within the food-banking culture and its nexus of relationships with donors that dampens the food banks' desire to empower the poor and take a more muscular, public stand against hunger.

THE MODERN FOOD BANK

A tour of one of America's state-of-the-art food banks is a striking experience, especially for one like myself who can remember the early hovels that passed for emergency food warehouses. Foodshare's president and CEO for almost twenty-five years, Gloria McAdam, showed me around its newest facility in June 2006—"newest" because Foodshare had outgrown four previous warehouses during her tenure. Built in the middle of a commercial industrial park that has consumed farmland where corn and tobacco once grew, the 30,000-square-foot building is a formidable structure that tripled the storage space of Foodshare's previous warehouse. Not only does its size and boxy shape conform to the other buildings along this commercial corridor, but it is no longer called an "emergency food warehouse" but a "food distribution center." The revised nomenclature puts Foodshare on a par with others in the food industry, a necessary condition of seeking substantial quantities of other people's food for free.

Although it may look like every other commercial building around it, Foodshare has incorporated a raft of energy-saving systems, devices, and gizmos, which has made it a candidate for Green Building designation. Coolers and freezers big enough to drive a semitrailer through have replaced the donated home refrigerators that provided a totally inadequate means of preserving food in the prehistoric days of food banking. A two-zone freezer holds regular frozen food in one area and ice cream in the other, which is kept at minus 10 degrees Fahrenheit. (There was enough Ben & Jerry's in stock that day to add five pounds to every Connecticut man, woman, and child.) The large cooler space is impressive not only for its size but also for what it says about the current direction of food bank inventory. About 54 percent of Foodshare's donated food is perishable, much of it fresh produce. While I saw shelves stocked to the three-story ceiling with high-sugar, high-fat products that the food industry deigned to donate, I also saw cases upon cases of fresh fruits and vegetables.

The emphasis on fresher, higher-quality food is as much a result of the changing structure of the U.S. food industry as it is the intent of food banking to provide healthier food. Computerized inventory control at major food manufacturers and retailers has resulted in less waste and spoilage, but it's also responsible for fewer donations of food to charities. Fortunately for food banks (but less so for their clients), the food industry still churns out something on the order of 15,000 new food products every year, many of which fail to meet their anticipated sales targets and consequently end up in the donation bin. But the overall impact of the industry's greater operating efficiencies has been less food, especially nonperishable food items, which were much easier for smaller, low-tech, local food charities to handle. For the more technologically endowed food banks, this shift in the food industry has meant that they must secure more perishable food to meet demand.

While food banks will do virtually anything to appease donors, the reverse is certainly not true. It is unlikely that any major for-profit food enterprise has made a significant business decision based on the needs of the emergency food-banking system. Increased efficiencies to reduce costs and in turn to satisfy stockholders have been accomplished through technology, consolidation, mergers and acquisitions, and simply moving or outsourcing to other regions or countries. Food banks in New Mexico, for instance, used to receive regular shipments of slightly dented cans, nonperishable food past code date, and other edible but nonsalable products. Not anymore. Major food manufacturers and wholesalers have found a secondary market in Mexico, which gives them a few pennies for each can of a product that can be sold in a country with lower safety standards.

Gloria McAdam shared the example of New England's largest supermarket chain, Stop & Shop, which used to operate regional warehouses, including one in New Haven, Connecticut. The proximity of this warehouse to Foodshare gave the food bank ready access to a steady supply of donated food. But in a cost-cutting move, Stop & Shop closed its regional warehouses, which were unionized, and contracted with C&S Wholesale Grocers out of Vermont, a non-union, lower-wage company. Stop & Shop reduced its costs, Foodshare (and other regional food banks) lost a major food donor, and workers received lower wages, forcing some of them to supplement their own food purchases with food bank donations.

But in spite of the ups and downs of the food industry, food banks have forged ahead with modernization and growth. Building facilities such as Foodshare's is an expensive proposition, made more costly over time as food banks are continually forced to expand to keep pace with the demand for food and more updated infrastructure. Foodshare completed a $4.5 million capital campaign to construct this facility, which includes enough additional land to accommodate another 20,000-square-foot expansion.

While Foodshare relies on about 2,000 individual volunteers each year to sort, count, clean, load, and distribute food, its 33 employees, from truck drivers to fundraisers, are the backbone of the organization. The 360 recipient organizations, which range from soup kitchens to church food pantries to nursing homes, received close to 9 million pounds of food from Foodshare in 2005, 80 percent more than they received in 1999. By one measure of food bank size—pounds of food distributed divided by the number of people in poverty who live in the region—9 million pounds of food places Foodshare among the top ten food banks in the country, a remarkable rating considering that metropolitan Hartford is smaller than other metropolitan areas. This rating suggests at least two things: the percentage of the region's people who are impoverished is very high (32 percent of Hartford's residents), and Foodshare has done an extraordinary job of soliciting and distributing food.

PROMOTING CODEPENDENCY AND MANAGING POVERTY

There are, in this rosy picture of a highly successful charity, a few troubling issues that take us back to our examination of the food gap. When I asked Gloria McAdam why the rate of food distribution growth was so steady and strong, she said that the "demand is bottomless; the more you provide, the more demand there is." In spite of Foodshare's role as a highly effective food bank—in fact, a model food bank in many respects—the demand seems to keep increasing.

Does the availability of food simply stimulate demand, or is demand itself rising? USDA statistics do suggest that Connecticut as a whole has made notable strides in reducing hunger and food insecurity over the past ten years. Among the seventeen states that have shown a drop in food insecurity (the rest have shown an increase), Connecticut ranks third. Among the sixteen states that have reduced the amount of very low food security (that is, hunger), Connecticut is second. Since the state's poverty levels have not changed significantly during this time, it is hard to know what other factors may have influenced this improvement. It may very well be that Foodshare, the state's other food banks, antihunger groups such as End Hunger Connecticut!, and the Hartford Food System have indeed done an exceptional job of mitigating hunger and managing poverty. But in spite of this apparent progress, more than 250,000 residents are still considered food insecure, and the conditions that produce food insecurity are as entrenched as ever.

What is inherently troubling about the relationship between demand/supply and hunger struck me one day when I observed a large Foodshare truck pulling into the parking lot of a north Hartford housing project. Since the residents knew in advance when and where the truck would arrive, they were already lined up at the parking lot's edge to receive handouts of food. Staff and volunteers set up folding tables and proceeded to fill them with produce, boxed cereal, and other food items offloaded from the truck. People remained quietly in line until it was their turn to receive a predetermined bag of food. There was no attempt to determine whether the recipients actually needed the food (although the project's high poverty level could be one proxy for a more formal assessment of need) and no attempt to encourage recipients to seek other forms of assistance, such as food stamps. The food distribution was an unequivocal act of faith based on generally accepted knowledge that this was a known area of need. The recipients seemed reasonably grateful in the way that those who have spent much of their lives waiting in line for something to be given to them typically are. More important, the staff and volunteers seemed even happier, having been fortified by the belief that their act of benevolence was at least mildly appreciated.

As word spread in the community, the lines got longer until the truck was empty. The following week, at the same time, the truck would return and the people would be waiting, only this time there would be more of them. It may very well be that a codependency between donor and recipient had developed. Both parties were trapped in an ever-expanding web of immediate gratification that offered no long-term hope of eventually achieving independence and self-reliance. As McAdam said, "The more you provide, the more demand there is."

Concern must be raised as well about the role that food banking's leadership plays in this evolving codependency and what might be best termed a stalemate in the battle against hunger. By leadership I don't necessarily mean the executive staff, but the board and officers of food banks, as well as those who volunteer their time and community reputations to head up the food banks' capital campaigns. These are the good citizens whose names and faces are prominently displayed when a food bank makes a big announcement, secures a major grant, or cuts the ribbon at the groundbreaking for a newer and bigger distribution center. They are generally professionals, often CEOs or other high-ranking officers of major corporations—in other words, stalwart members of the community with a sense of noblesse oblige. Rarely will you find a person of color or someone who has received assistance from a food bank among them. They are people of power and influence, who are using both to serve their community and the needy.

Food banks are among the largest and most influential charities in most communities, rivaling hospitals, the United Way, and YMCAs/YWCAs. Since influential people are drawn to influential organizations, both food banks and the people who run them are in a unique position to promote a vital public discourse around hunger, food insecurity, and poverty. Do they? Generally speaking, they do not, because influential people don't attain exalted positions within a community's hierarchy by asking hard, controversial questions or by becoming agitators. Upsetting the apple cart is not the way it's done in polite society. Food banks will be unable to curry favor (receive money and food) if they offend corporations; the corporate leaders on food bank boards are not going to demand that state legislatures and Congress increase their taxes so that government will have more resources with which to tackle hunger. The codependency that exists in that north Hartford housing project parking lot is mirrored in corporate and charity boardrooms across America.

In *Sweet Charity?* Janet Poppendieck draws a startling conclusion about the complex role that food banks play in contemporary American society:

> What I have found in seven years of studying the growth and institutionalization of the emergency food system is that emergency food has become very useful indeed . . . The United States Department of Agriculture uses it to reduce the accumulation of . . . agricultural surpluses. Business uses it to dispose of . . . unwanted product, to . . . avoid dump fees, . . . and to accrue tax savings . . . Churches use it to express their concern for the least of their brethren . . . Environmentalists use it to reduce the solid waste stream . . . A wide array of groups, organizations and institutions benefits from the halo effect of "feeding the hungry." If we didn't have hunger, we'd have to invent it.

Yet the question she asks, one that has been asked increasingly by many people over the past ten years, is, what could the effect of food banks be if all the energy that was put into soliciting and distributing wasted food was put into ending hunger and poverty? Surely, two thousand Foodshare volunteers, led by some of Connecticut's most privileged people, showing up one day at the state legislature demanding enough resources to end hunger, would have a sizable impact. Multiply those volunteers by three or four times—the number of volunteers in the state's other food banks and hundreds of emergency food sites—and you would have enough people to dismantle the Connecticut state capitol brick by brick and move it to Pittsburgh. Put all the emergency food volunteers, staff, and board members from across the country on buses to Washington, D.C., to tell Congress to end hunger, and you would have a convoy that would stretch from New York City to our nation's capital.

The good news is that food banks are beginning to take public policy approaches more seriously. They are also engaging in a variety of other activities and projects that are expanding their repertoire of hunger-ending solutions. The bad news is that these strategies are still not enough. Even so, some of the promising new strategies deserve notice, as well as consideration for how they might become more extensive.

NEW DIRECTIONS FOR FOOD BANKING IN NEW MEXICO, MASSACHUSETTS, WASHINGTON, D.C., AND OREGON

At the Storehouse food pantry in Albuquerque, New Mexico, that state's largest emergency feeding site, the food line starts forming at 7:30 in the morning. Mostly women, many small children, and some single men are shaking off the morning chill, hoping to be one of the first one hundred people let into the food pantry. Just fifty yards off Albuquerque's historic Route 66, the Storehouse has only enough food to feed one hundred families per day, which means that if you don't make it in, you'll have to try again the next day or the day after that. People are turned away every day. It's not that the Storehouse's staff and one hundred volunteers don't want to serve everybody who's hungry; it's just that the need won't stop.

"In 1999, we served the equivalent of 200,000 meals each year," said Lee Maynard, the Storehouse's executive director. "Right now [2005] we're serving 1.4 million meals a year, which is 45 percent more than we served the year before. Things are getting worse." While the lines and the numbers suggest an air of desperation, once inside the Storehouse, you feel the gentle touch of a caring place. Hungry people are called "customers" and escorted through aisles similar in many respects to those of a small supermarket. Unlike most emergency food sites around the country, the Storehouse uses what's called a "customer choice" (also called "client choice" by some sites) model that allows people to select their own food from various categories. The amount of food is based on the number of people in a household and is designed to give each shopper twenty-one meals per household member (one week's worth).

Why do people come to the Storehouse? Shifts in the economy away from higher-paying manufacturing jobs to low-paying service jobs are forcing more people to seek assistance, both from private charities such as the Storehouse and public agencies such as the Food Stamp Program. One New Mexico county food stamp director told me that he was seeing an increase in the demand for food stamps, a situation that was common across the state. What was surprising was his reason for the increase. In his opinion, there were ample employment opportunities in the region, but because the wages and benefits at two newly opened Wal-Marts, now the county's biggest employer, were so low, their employees didn't have enough money to buy food. Because Wal-Mart's employees were eligible and applying for food stamps, the federal government (that is, taxpayers) was, in effect, subsidizing Wal-Mart's operations.

Even though Lee Maynard's service treats needy people with as much respect and dignity as possible, it is, in his opinion, a service that is only treating the symptom of a much bigger problem. "We'd be happy to never see our customers again," he confessed, "and they would be happy to never see us again." But without significant structural changes in the economy, it is not likely that programs such as the Storehouse will go out of business anytime soon.

The "client/customer choice" food distribution model and similar improvements in the care and treatment of needy people have—along with the food that is available, of course—eased the pain and indignity of poverty. But like the rotten potatoes I was forced to receive with obsequious gratitude, food banks still must take whatever they are given, which means that the quality and selection can vary enormously. On the day that I visited the Storehouse, there was an ample supply of nonperishable food products, but meats, dairy products, and especially produce were barely adequate. A bin of iceberg lettuce had as many brown leaves as pale green ones, and a large cardboard carton of cantaloupes was only hours away from the compost pile. In general, food banks have become increasingly aware of their responsibility to both their recipients and their donors and have been utilizing a number of strategies to improve the quality of the food they offer.

The Food Bank of Western Massachusetts, located in Hatfield, is a prime example of an emergency feeding institution that has linked food distribution to the needy with the production of the healthiest food available. Rather than be dependent on the ebb and flow of food donations, the Food Bank decided to "grow their own." Over the course of almost twenty years, they acquired sixty acres of prime Connecticut River bottomland and developed a community supported agriculture (CSA) farm that has more than six hundred shareholders, who provide the bulk of the farm's financial support. The shareholders receive about half of the produce, while the other half goes to the Food Bank for distribution to four hundred local food pantries.

Even though the farm produces an exceptional quantity of high-quality organic produce (70,000 pounds of winter squash were harvested and given to the food bank in one week alone), it doesn't fully engage the shareholders (nonrecipients) or the recipients in ending hunger and food insecurity. According to David Sharken, former executive director of the Food Bank, the farm "builds community, but does not necessarily ensure food security for low-income populations . . . The project is not creating food self-reliance among low-income families." And like other food banks, this one is running at full speed just to stay even with demand, including going through its own capital campaign in order to expand its facilities.

Like the Food Bank of Western Massachusetts, the Capital Area Food Bank in Washington, D.C., has access to a farm to grow produce for low-income Washingtonians. The farm is on thirty acres of land that is part of a much larger parcel owned by the Chesapeake Land Trust. The Capital Area Food Bank also employs a CSA model that divides its produce between itself and higher-income shareholders, but in addition it sells a large share of its produce at farm stands and farmers' markets in areas of the District of Columbia that have very few high-quality food stores. In effect, this food bank uses the farm to close one part of the food gap in areas where both farmers' markets and retail food stores have failed. It also employs public policy approaches by using the federally funded Farmers' Market Nutrition Program (FMNP), which gives low-income families special coupons to buy fresh local produce. In other words, the Capital Area Food Bank is "growing their own," but bolstered by public policy support, it's also using its produce and charitable efforts to stimulate economic activity that will increase the access of residents to healthy and affordable food.

Lack of a meaningful connection to public policy has caused many food banks to fall short. Challenging the powers that be is very difficult when a food bank's board of directors comprises many well-heeled conservatives. A notable exception is the Oregon Food Bank, based in Portland. Looking for ways to have a more systemic impact on hunger and its root causes, CEO Rachel Bristol began to move the organization in the direction of public policy advocacy. It took ten years, but the food bank's board of directors now has an advocacy committee. Its budget also reflects the shift in philosophy by paying three full-time people to work on advocacy. In the past, it had only one half-time person in this role.

The Oregon Food Bank has addressed the root causes of hunger by addressing the need to increase the minimum wage and encouraging low-income food bank recipients to take advantage of the Earned Income Tax Credit. The food bank has taken a strong position in support of expanding the Food Stamp Program and, closer to home, has spoken out on Oregon ballot measures that affect the state budget and taxes. While some food banks have hired lobbyists to secure state funding to, in effect, enable them to buy more food, the Oregon Food Bank has used a lobbyist to address state welfare legislation that would directly benefit low-income families. According to the food bank's director of advocacy, Kim Thomas, "We'd like to see more focus nationally on income support programs and some national statements about growing income inequality, which really is the root cause of hunger."

There has been pushback from some of the food bank's food industry donors that have found the idea of a higher minimum wage a little hard to swallow. The food bank listened to what they had to say and, in the end, held firmly to their position. In fact, one donor even said, "We're so glad to be giving money to an organization that isn't just moving food."

The Oregon Food Bank admits that it still measures its performance in food poundage, but it is increasingly committed to finding long-term solutions to hunger and food insecurity. The leaders believe that those solutions will be found in the policy arena, not in a bigger food warehouse or more efficient methods of managing waste. And they may be right: Oregon went from having the highest level of food insecurity in the country to being twenty-first in only three years (2002–2005). It had the fourth-highest decrease in food insecurity and the highest decrease in very low food security. This remarkable progress came about as a result of working together and focusing on policy issues.

THE CURRENT LANDSCAPE

CHAPTER SIX

Re-Storing America's Food Deserts

Left to themselves, economic forces do not work out
for the best except perhaps for the most powerful.
John Kenneth Galbraith

THE FEBRUARY 1996 ISSUE OF *Connecticut* magazine proudly proclaimed: "The food fight is coming on strong in Connecticut, where supermarkets are adding almost every service imaginable. Food courts, dry cleaners and coffee bars are showing up in stores from Bristol to Bridgeport, Manchester to Monroe... Cutting-edge stores such as Big Y and Shaw's now offer open-air bakeries and see-it-all butcher shops, while others allow you to load up on everything from head cheese and halogen lamps to Cheerios and stereos."

The piece continued in this puffy prose style for several pages, extolling the excitement of the sexy new food emporiums popping up across the state faster than mushrooms in an August forest. "Springfield-based [Massachusetts] Big Y, which opened three Connecticut stores last year, plans to add seven more this year. Shaw's, based in East Bridgewater [Massachusetts] was not in the state in 1994, but by the end of 1996 will have at least eight Connecticut stores. Stop & Shop [also based in Massachusetts] is turning its traditional stores into massive Super Stop & Shops filled with 52,000 items." The competition was intense and the investments were serious—$2 million to $3 million per single-store renovation and $4 million to $13 million per new store. Anyone getting in the way of this brave new world of Connecticut food shopping would be roadkill.

In the same month and year that *Connecticut* was waxing ecstatic over the dazzling array of products now available at the state's ever-growing list of supermarkets, a team of four Yale MBA students released a study for New Haven's Greater Dwight Development Corporation (GDDC). In more controlled MBA style, the students reported, "With only one supermarket on the outskirts of the city, New Haven needs a major supermarket in its inner-city. Prices at grocery stores within New Haven are 6 percent to 21 percent higher than prices at suburban supermarkets. [These] suburban supermarkets have more product selection and better quality produce when compared to [New Haven] grocery stores. [And] many households within New Haven do not have access to a car . . . and go to the trouble and incur the cost of taking a taxi or using public transportation to travel to a suburban supermarket." The report went on to detail plans for the development of a new supermarket chain store in one of the city's poorest neighborhoods, a place that the graduate students' research had concluded held sufficient buying power to support such a store.

While Connecticut's suburban soccer moms only had to worry about whether to purchase the domestic or the imported Brie, low-income urban residents had to choose between a $10, thirty-minute taxi ride to the suburban supermarket and their neighborhood store's overpriced, wilted lettuce. Not a single inner-city area in Connecticut's three major cities—New Haven, Hartford, and Bridgeport—were proposed sites for the much-touted supermarket expansion. The only exceptions to this food retail industry pattern of avoiding the cities and building in the affluent suburbs were a couple of chain stores that effectively straddled a city-suburban boundary, sites that were no more accessible to inner-city residents than were the suburban stores.

WHY DON'T SUPERMARKETS SERVE LOW-INCOME AREAS?

Well over 80 percent of Americans buy most of their food in full-size supermarkets, mostly large chain supermarkets. In 1968, Hartford had thirteen chain supermarkets operating within its city limits. Shortly after the civil disturbances of that year and the resulting population shift, the stores began the process of closing, pulling up stakes and relocating to the suburbs. At the time I commenced my tenure at the Hartford Food System in 1979, only six stores were still open. (By 1986, there would be only two.) You might say they followed their more affluent shoppers to the suburbs. It was also true that the supermarket industry was rapidly moving to a much larger store format that would simply not fit the small, cramped spaces of densely populated urban neighborhoods. You might also say—without a lot of direct proof but with a good deal of intuition—that the traditional food retail industry did not want to serve a predominantly nonwhite, lower-income shopper base.

A more concise business explanation of why chain supermarkets don't adequately serve lower-income urban communities was offered by the Hartford Food System. In a 2006 report, the organization said:

> Both median income and population density have a close relationship to the amount of supermarket capacity that can be found in a given community, with the former bearing a somewhat stronger correlation to the pattern of store locations than the latter. These two factors help explain grocery retail gaps in Connecticut... Like any other industry, supermarkets are in business to make money. Chains usually build stores in places where the profit-making potential is the greatest and the financial risks are most favorable to their company... Modern supermarkets are thinly margined businesses that require enormous sales volume to make a profit... Supermarkets primarily base their location decisions on the revenue projections and number of targeted customers they can reach within the trade area... In some economically distressed areas, chains would be reluctant to open a store [because] the anticipated sales volume just wouldn't be enough to support a full-size supermarket.

It would be hard to make an outright charge of discrimination, or redlining, on the part of the supermarket industry toward low-income communities of color stick. But as the Hartford Food System made clear, the retail food industry goes where it can make the most money, and those are not places that are heavily populated by low-income African American and Hispanic households. Supermarkets are also not charities, nor can they take tax write-offs for acts of goodwill and social responsibility. Even so, the business decisions they make are often disadvantageous to lower-income communities in both rural and urban areas.

A national study of the nation's largest metropolitan areas by the University of Connecticut Food Marketing Policy Center makes the results of our capitalist business model clear. In Connecticut's poorest urban communities—Hartford, New Haven, and Bridgeport—there is about 1.6 square feet of supermarket space for every resident, compared to between 5 and 7 square feet in the relatively affluent and nearby suburbs—Ansonia, Wallingford, and Rocky Hill. On a nationwide basis, the policy center found that zip codes with the highest percentages of households on public assistance had less supermarket space per capita than higher-income zip codes. A study of food stores in Maryland, Minnesota, Mississippi, and North Carolina found that chain supermarkets were four times more likely to be located in predominantly white neighborhoods than in black neighborhoods.

Again, the supermarket industry's explanation makes good business sense. Its spokespersons claim that the operating expenses of inner-city supermarkets, including rent, insurance, and security, are higher than those of non-inner-city stores. They also acknowledge that they have moved to a cookie-cutter, one-size-fits-all approach to new store development. For efficiency's sake, they need to build larger stores that all look alike and are configured in the same way. This means the oddball-size stores that used to exist in inner-city locations, with one small loading dock that could not accommodate an eighteen-wheeler, do not fit the plans for corporate expansion. Since densely built urban areas do not have sufficient land to accommodate the larger stores, which need huge parking lots and ample turning space for large trucks, new stores are rarely built in cities.

What has been the impact on those who haven't been able to flee the cities for the suburbs or don't own a car? In the early 1980s, 22 percent of Hartford's residents lived below the national poverty level. That number would grow to nearly 32 percent by the 2000 U.S. Census, thereby making Hartford the second-poorest city in the country. Based on Bureau of Labor Statistics household purchasing data, a low-income family in 1982 spent 25 percent of its income on food that it consumed at home (as opposed to additional food the family might purchase at a restaurant). A middle-income family spent 15 percent of its income for the same purpose, and an upper-income family spent 6 percent of its income.

The Hartford Food System and the Citizen Research Education Network conducted a forty-four-store food price survey in 1983. It found that city supermarkets, as I had observed personally upon moving to Hartford, were between 14 percent and 37 percent more expensive than comparable suburban stores. If a family of four bought all of its food in Hartford stores, which later research found was the case for 25 percent of the city's low-income residents, it would spend up to $1,500 per year more than a family that shopped elsewhere. In 1983, the poverty level for a four-person household was $9,900. Shopping in Hartford would push the family's annual food costs to more than one-third of its income. That the poor would pay more and be forced to devote a much larger share of their income to food, or simply buy and eat less, was perhaps the most striking feature of the food gap at that time.

By the 1990s, it had become a matter of canon that the poor were badly served by America's supermarket industry. In "Higher Prices, Fewer Choices," a 1990 report produced by Public Voice for Food and Health Policy, the problem of limited access to affordable food stores in America's rural areas was documented as well. Poor access to retail food stores was not just an urban phenomenon as people began to discover its country cousin, the rural food gap. Nationwide, not only were the poor paying more for food and selecting from lower-quality products, but the lack of access was beginning to affect their health and their ability to use their food assistance dollars effectively. At a 1995 conference designed to highlight the supermarket gap in poor areas, USDA secretary Dan Glickman said, "Restricted or limited [food] access undermines the [USDA's] ability to promote health through nutrition, because if prices are too high, there is not enough bang for the buck for Food Stamps and WIC (Women, Infants, and Children Program), or if choices are limited . . . [Americans] can't make the choices that nutrition education efforts encourage them to make." In other words, we were wasting time teaching people to eat healthfully if they couldn't find affordable food stores, and they were not getting the full value of the food assistance vouchers that taxpayers were supporting.

ENTERING THE RETAIL FOOD STORE BUSINESS

By the early 1980s, market forces had swept Hartford clean of nearly all its supermarkets. The need was still there, however, and the market research strongly suggested that Hartford's neighborhoods could support supermarkets. In 1984, the Hartford Food System decided to enter the world of community economic development, which, simply put, marries social goals with economic and business goals. With respect to food, community economic development strategies require that nonprofit organizations enter the marketplace, run a food enterprise in a businesslike way, and provide for as much community participation and benefit as possible. To continue with the marriage metaphor, if a businessperson married a social worker, their offspring would likely grow up to be community economic development practitioners.

The bands were playing and the people were dancing on Hartford's Farmington Avenue one beautiful September day in 1985. The reason for this public display of merriment was the opening of a brand-new supermarket with the name Our Store, a cooperative that was owned by 1,200 neighborhood residents who had each invested $50 of their own money. Most of the residents were from the lower-middle-income neighborhoods of Asylum Hill and the West End, only a frog's leap from Mark Twain's former home. The new store's name said it all: this was a store that residents had started themselves (along with a total outside investment of $500,000 in corporate social responsibility and City of Hartford funds), so they had every right to be twisting in the street. Not only were they the investors, but they also controlled the business because as a cooperative, each member—no matter how much he or she invested—had one vote.

Many of the dancers were elderly men and women, who shook their hips slowly but gracefully. They pivoted their bodies gingerly around their canes as if the canes were dance partners and occasionally tempted fate by shaking their canes high in the air. The co-op's board of directors were laughing, shaking and swaying, and chatting with city officials, lenders, investors, and the press. Joe DiBattista, Our Store's manager, was nattily dressed in a new suit he had bought for the occasion and was giving everyone a cheery greeting as they entered the completely renovated store.

What happened that day in the parking lot of that restored, reopened, and renamed A&P Supermarket was no garden-variety celebration. The A&P, the last one in the city and one of the few left in the Hartford region, had closed for good two years earlier, stranding the neighborhood's residents. The reopening and its accompanying celebration were, in effect, a collective sigh of relief, born of frustration and hope, that this community, through its own blood, sweat, and equity, had turned the tide against a wave of supermarket abandonment. The residents had been left, like so many other economically distressed communities across the nation, high and dry in a food desert. They had taken their food future into their own hands, and for the moment, as the gleaming new façade, high-waxed linoleum floors, and jam-packed grocery shelves attested, they had won the battle against the supermarket industry's willful neglect.

Why did ordinary citizens and the Hartford Food System, neither of which had any formal training in supermarket development and store management, suddenly take matters into their own hands? The fact of the matter was that none of those people were dying to start and run a 13,000-square-foot co-op food store. Most of them would have been happy to shop at the older 1950s-era A&P and First National stores that had dotted the city for much of the past forty years. But now those stores were boarded-up, graffiti-covered shells that were only waiting in line for the wrecking ball. The passion that was moving people to dance, make speeches, and slap one another other on the back was the flip side of the anger and frustration they had felt for so many years.

People had tried many methods to stop the corporate food chains from shuttering their stores. In the early 1980s, when it was rumored that one small chain store located in a working-class Hartford neighborhood might close, residents used a technique from the celebrated community organizer Saul Alinsky. A small crowd arrived at the suburban home of the chain's president at three o'clock in the morning, banging pots and pans. They got his attention—that store never closed, although it was eventually sold to a smaller chain, which never reinvested in the site and allowed the store to deteriorate badly. In the city's West End, known as the most civil of Hartford's neighborhoods because of its high concentration of professional households, the closing of a favorite natural food store called Cheese 'n Stuff stirred angry protests and demands that city hall do something. The store was bought by the natural food chain Wild Oats Markets, which promised not to close the store. But when Wild Oats opened a store in the nearby affluent suburb of West Hartford, it reneged on its promise and closed Cheese 'n Stuff.

That something as mundane as a grocery store can provoke such a range of emotions speaks to our sense of community as well as our basic need to eat. Anger, hope, frustration, and giddy optimism have driven the local food store debate for the past thirty years. Sociologists and community economic development activists tell us that a lot of a community's sense of self-worth and social health depends on the presence of a supermarket. The economic strength of a neighborhood may be measured in part by the existence of a nearby, high-quality, well-maintained food store. Such a store assures residents that theirs is a stable, safe, and viable place to live and work. And like other businesses, supermarkets pay taxes and employ workers—up to 250 people for the larger superstores.

Lack of access to affordable supermarkets is rising to the top of the list of public health concerns as well. One recent study found that a person's ability to achieve a healthy diet is influenced by the availability of food stores. Conversely, a person is more likely to have an unhealthy diet if he or she has easy access to fast-food restaurants and other similar food service establishments. This finding correlates with the fact that lower-income, racially segregated neighborhoods have fewer healthy food choices and higher rates of diet-related disease. This contention is backed up by the USDA's Economic Research Service, whose research has shown that low-income households in rural areas and poor central cities have less access to reasonably priced, high-quality food than other households.

One of the most interesting studies sponsored by the Hartford Food System, "Food Security and Community" by Katie S. Martin, found a strong relationship between food insecurity and a household's ability to travel to a large supermarket and shop frequently. In other words, if a person can get to a supermarket and shop when he or she needs to, he or she is less likely to be hungry or experience other characteristics of food insecurity, even if the person is low-income. Of the 330 households surveyed, all of which were below 185 percent of the federal poverty level, 52 percent were deemed food insecure or hungry using the USDA's household hunger and food insecurity survey.

In 1987, almost nineteen months after the bands stopped playing and the members put away their dancing shoes, the Our Store co-op closed its doors and covered its windows with plywood. As well-intentioned as the effort was, the enthusiasm and good vibrations were not enough to overcome the harsh realities of a marketplace that demanded 60,000-square-foot superstores crammed full of every item under the sun. The Hartford Food System, the co-op's members, and the many able and talented partners who had rallied around the valiant little enterprise were not supermarket managers or financial wizards. They were social entrepreneurs who knew how to organize people and resources, but they hadn't been brought up in the rough-and-tumble world of food retailing. One of the biggest mistakes they made, for instance, was to secure an insufficient amount of capital. This failure prevented the co-op from having enough cash to hire the talented managers and staff necessary to make a small store in a challenging marketplace zing. There were many other factors in its demise, including a hostile landlord, a badly negotiated lease, and inexperienced amateurs like myself, who, among other things, accidentally destroyed the store's $300 vacuum cleaner while cleaning the produce cases.

There is no guarantee that other similar co-op enterprises will prosper with adequate financing and a higher level of business acumen, especially where lower-income households are the major customer base. Many attempts have been made and most have failed, yet the attempts continue. When food co-op stores have succeeded, it has been primarily because they have appealed to a higher-end, well-educated member/customer base with a highly evolved food consciousness. The Park Slope Food Coop in Brooklyn, New York, and La Montañita Co-op in Santa Fe, New Mexico, are two examples that come to mind. But the most frequently stated desire in lower-income communities is to have reasonably good access to an up-to-date chain supermarket. It is not to invest in co-op–style stores.

THE FOOD DESERT SPROUTS NEW LIFE

My untutored observations of nature suggest that when a landscape is decimated by either a natural or a man-made disaster such as a forest fire or clear-cutting, biology usually responds quickly and logically to fill the void. Small, though not necessarily desirable, plants take over, resulting in rather hodgepodge and scraggly growth patterns. While aesthetically unpleasing, these patterns and their accompanying mishmash of flora create, over successive generations, a new and vibrant habitat.

A drive through Hartford's neighborhoods today is not unlike a visit to a previously ravaged landscape that has had twenty years to heal itself. The first impression is that this is a place bereft of all but a few motley mom-and-pop stores pushing tobacco, alcohol, fat, and sugar. A closer inspection, however, reveals that the laws of the marketplace have a rough kind of street justice, not unlike that of a natural environment attempting to conceal its scars. Retail food stores have partially filled the gap left by the loss of larger supermarkets in a uniquely urban, slightly gritty, and crudely functional way. They have altered the food retailing environment for the better compared to what it was twenty years ago, but when it comes to price, quality, and selection, they do not compare to the shiny supermarket cathedrals of the suburbs.

A drive down Franklin Avenue in what is referred to as Little Italy must include a stop at D&D Market. Located in a clean and pleasant mini-mall, it's *the* place for both real Italians and Italian wannabes to buy olives, fresh mozzarella, spicy meats, and Arthur Avenue bread. The store has staked out a niche that is unassailable by anyone who might have the nerve to enter the specialized (and by all accounts secretive) world of Italian food retailing. Very small elderly women who speak only Italian approach the deli counter with the assertiveness of an NFL linebacker, demanding samples of prosciutto and salami from the two-hundred-pound clerk. A request ignored results in a withering fire of invective accompanied by a carefully orchestrated set of hand gestures.

Farther north, a stroll down Park Street takes you into the heart of the city's Hispanic community. El Mercado is an imitation of a small-town Puerto Rican marketplace where you can find piles of rice and beans, Goya brand packaged goods, and incredibly fresh cilantro. Spanish is the dominant language and influence, but a couple of Asian takeout places and pizza shops have sprouted in the mix, giving the place the eclectic flare of something either uniquely cosmopolitan or woefully uncertain of its identity. El Mercado was built with a lot of public cash and soft money (grants or loans that nobody expected to be repaid) in the early 1990s, and while its origins and authenticity may be questioned, it nevertheless stands as a slightly tattered emblem of the community's Caribbean culture.

As you wend your way north up Blue Hills Avenue to the city's northern border, you'll find Model Market. The 5,000-square-foot store was actually "reopened" in 2002 by business partners Edwin Serrata and Ruddy Fernandez after it had been closed for almost five years by a fire. Violet neon lights frame the store's large bay windows, giving it a nightclub ambiance. Model Market, like most of the larger grocery stores in the city, accepts Electronic Benefit Transfer (EBT) cards, which are used by food stamp recipients to purchase food. The back of the store houses the Butcher Chop, whose products may be bad for the human heart but are the heart of the business. Like many groceries in low- to middle-income ethnic neighborhoods, meats are far and away the biggest sellers. A good meat section is still more important in most urban areas than a good produce section, although Model Market tries to bring a little more dietary balance to the community with its signature fruit gift baskets.

Model Market rose from the ashes through sheer force of neighborhood will. As is the case in many urban neighborhoods across America, the residents of Blue Hills simply refused to go storeless. According to Glenn Geathers, who in 2002 was the project manager for the Hartford Economic Development Commission, the reopening of Model Market "was a priority of the Blue Hills neighborhood." That priority meant that a modest amount of public financing was made available for the start-up, which was enough to help experienced food store operators Serrata and Fernandez transform the urban food desert into a modest oasis for low- and moderate-income shoppers.

Traveling southwest from the Blue Hills neighborhood, you find Unity Plaza, one of the city's earliest and most beleaguered attempts to infuse an impoverished community with some commercial vitality. Located not far from what used to be (until it was bulldozed) Hartford's most notorious housing project, the plaza is the site of C-Town Supermarket, owned and operated by Jeffery Diaz. Diaz has seventeen years of grocery store experience, much of it gained at his parents' knees in their Long Island City, New York, store, which they opened in the 1980s after their arrival from the Dominican Republic. He opened the C-Town Supermarket in 2001, in a place that is best known for failed food stores, tough-as-nails customers, and constant street crime. In what might be considered in gentler and kinder environs an unnecessary display of machismo, Diaz says that the Unity Plaza location pales in comparison to his former Long Island City store, where bullet holes in the wall were regarded as marks of distinction.

When Diaz first opened the Unity Plaza store, shoplifting was rampant. But he brought the hammer down hard and prosecuted everyone he caught. Since the neighborhood's initial testing period, Diaz claims that he hasn't lost a pack of gum. Although he confesses that some food distributors refuse to deliver to his store because they think that the neighborhood is unsafe, he insists that business is good and getting better. "Dominicans like hard, physical work," he says, "and this job requires hard work." That's at least sixty hours a week for Diaz.

In addition to public money and tough-minded entrepreneurship, there is one other factor responsible for the city's modest revival in urban food retailing. It's the wholesale distribution network, without which no retail food store could survive. As supermarket chains get larger, increase the number of stores, and consolidate the number of chain store corporations, they are able to demand ever greater concessions from their wholesale suppliers. For these suppliers to meet those demands, they in turn get bigger, seek greater efficiencies from larger-scale operations, and cut their costs. There is simply no place in this system for a single, inner-city grocery store of 15,000 square feet. The scale of the modern wholesale/megastore distribution chain challenges the imagination and far surpasses anything that modest retail operations could ever hope to gain access to.

To fill this void, medium-size, member-owned distributors, typified by Krasdale Foods, a New York City–based wholesaler, have emerged to serve urban food stores such as C-Town and Model Market. Krasdale is a virtual co-op that combines the buying power of many smaller food stores. If you combined all of Hartford's seven independently run grocery stores into a single retail unit, it would fit comfortably inside one suburban supermarket. The urban independents simply aren't large enough on their own to command the kind of service and price breaks that large chains get from major food distributors. Without Krasdale, Hartford's grocery entrepreneurs could not buy groceries at a reasonable wholesale price, receive inventory financing assistance, or get advice on advertising, merchandising, and a host of other operational details. While Krasdale doesn't come close to doing what C&S Wholesale Grocers does for Stop & Shop and other major chain supermarkets, it provides a necessary lifeline for small city stores, which in turn provide a lifeline for low-income city shoppers.

Hope springs eternal in the breast of Glenn Geathers, who is convinced that a major chain store can be brought back to Hartford. As one of the city's most optimistic guys, Geathers cut his development teeth with the Hartford Tenants Rights Federation in the city's Northend, where he worked like a dog to bring a new supermarket to one of Connecticut's poorest neighborhoods.

"No is only a precursor to yes," Geathers said while shrugging off the slew of rejection letters he's received from developers and supermarket chains. He readily admitted, however, that in the eyes of those developers, Hartford "doesn't have the rooftops to support a major chain store." "Rooftops" is developer talk for insufficient purchasing power due to the city's lower-density residential profile and its high concentration of low-income households. But the negative assessment of the city as a marketplace for new supermarket construction didn't stop Geathers. Because of his persistence over the course of several years, the Clay/Arsenal neighborhood secured control of an appropriate site on North Main Street, a commitment of $5 million in public and private financing, and strong interest from several store operators. Eventually, Geathers succeeded—sort of. He got a shopping center anchored by a small chain store built in the neighborhood. It opened in 2005 and is known for low-priced nonperishable goods and a mediocre assortment of perishable items. Are the residents of this hardscrabble community better off than they were five years ago? Absolutely. Are they as well served as the residents in the surrounding suburbs? Not on your life. But in a community that suffers three setbacks for every one advance, the incremental gain represented by this new store is a victory worth celebrating.

APPLES, BEANS, AND COLLARD GREENS: A ROCHESTER, NEW YORK, NEIGHBORHOOD FIGHTS BACK

Hank Herrera is a colleague and fellow community food activist from Rochester, New York. He told me the story of why he became interested in food issues, an interest that he has pursued for fifteen years now. One day he was driving around the Upper Falls neighborhood, one of Rochester's lower-income communities, in the dead of one of the region's notoriously severe winters. He spotted an elderly woman, bundled up against the blistering wind, walking across a bridge. She was carrying a bag of groceries in each hand, taking slow, mincing steps across patches of treacherous ice. She had purchased her food at a grocery store in an adjoining middle-class neighborhood and was completing her roundtrip journey of more than two miles. "I couldn't get this image out of my head," Herrera told me, "of an old woman enduring so much hardship just to get her food."

As a person who's better known for taking action than merely feeling guilty, Herrera joined up with a neighborhood group called Partners Through Food (PTF), which had decided to address the community's lack of a full-line supermarket. The Upper Falls neighborhood had a median household income of $11,000 in 1990, making it the third poorest of Rochester's thirty-five neighborhoods. What was worse, especially for food shoppers, was that half of the community's members did not own a car. Public bus routes required at least one transfer to reach downtown Rochester, where most neighborhood residents were forced to shop. By

1992, after a series of store closings, only one small, family-owned supermarket served the neighborhood, along with a growing number of convenience stores that offered only expensive packaged and canned food.

PTF sought and received assistance from the City of Rochester in the form of a $25,000 grant to conduct a market study. The group reached out to community members through surveys and meetings to assess their needs and desires. And it took up a series of critical questions regarding supermarket development and operation: Should PTF try to be a new store's owner/operator, in the mode of a food co-op? Should it be the store's developer but leave the operation and ownership to an outside supermarket firm? Or should it simply play an advocacy and agitator role to try to get the city and the supermarket industry to respond to the neighborhood's needs?

These questions and how they were debated guided PTF's journey over the next three years through a series of twists, turns, and dead ends. The group worked on purchasing a several-acre site that included a modest-size, closed supermarket. It prepared a business plan with the help of an outside supermarket consultant. It negotiated with various potential operators and competed for site control with other potential operators who did not intend to develop the kind of store residents wanted. Group members marched on city hall, chanting "Apples, beans, and collard greens" to make it abundantly clear to everyone what the neighborhood needed.

Things took a turn for the worse in 1995 when one potential operator for the site pulled out and the supermarket in downtown Rochester, where Upper Falls residents were shopping, closed. This particular chain had been solicited to develop and operate a store in Upper Falls but had declined, saying that it's marketing strategy was to operate suburban superstores. At this point, PTF's members were feeling exhausted and dispirited by their attempts to bring in a supermarket and to assume some kind of ownership and operational role. They realized, according to Herrera, that "what the residents really wanted was a decent, full-service supermarket large enough to offer a decent selection of food. They did not have a significant interest in ownership or in regional food production and distribution issues."

Surprisingly, the City of Rochester decided to enter the fray. Under the leadership of the city's first African American mayor, William A. Johnson Jr., the city raised economic development funds and began negotiations with a number of supermarket chains. In September 1996, the mayor called a meeting of PTF board members at which he announced that the Buffalo-based Tops supermarket chain had agreed to open not one but five supermarkets in Rochester, including one in the Upper Falls neighborhood. Three of the five stores would open in city neighborhoods that had no supermarkets, and two would involve substantial renovation and upgrading of existing facilities. The total development package was worth $36 million, including $23 million from Tops and the balance from a variety of public sources. As a bonus, Tops agreed to hire 80 percent of the employees for the new stores from their respective neighborhoods.

A beautiful new supermarket opened in Upper Falls on April 1, 1998, with PTF board members and neighborhood residents leading a parade through the streets to the brand-new supermarket plaza. PTF's banner proclaimed the source of their strength and commitment: "We Have Come This Far by Faith." Since that time, the shopping center has been fully occupied, and food sales at the Tops are brisk. Unlike the Our Store experience in Hartford—owned and operated by its members—the Tops supermarket was owned and operated by a private chain, whose deep financial pockets allowed it to hire superior management staff. Hartford chose community ownership and failed. Rochester intentionally sacrificed community ownership, kept the pressure on the powers that be to find a solution, and secured a successful grocery store that met their needs for high-quality, affordable food and a substantial number of community jobs.

Of course, there is a risk associated with this approach. By sacrificing any significant measure of business control, the community placed itself at the mercy of a major supermarket chain, a place where hundreds of communities have been in the past and have lost. Yet the community has what it wanted most—a good supermarket—and as time goes on and market forces shift again, as they inevitably will, the community must be prepared to continue to fight.

NEW HAVEN TAKES A ROAD MORE TRAVELED

While the supermarket industry is only now waking up to the potential to "re-store" underserved communities—at least in urban areas where restoration and gentrification are generating new wealth and buying power—the story for the past twenty-five years has been about community efforts, such as those in Hartford and Rochester, to close the food gap. As the two previous stories illustrate, food store restoration work has required a combination of sophisticated community development strategies, often in tandem with community organizing and advocacy efforts, supported by public and sometimes philanthropic investment. The supermarket story from New Haven, Connecticut, presents what is now fairly standard practice in the world of commercial retail food development.

Until the summer of 1998, residents of New Haven's Greater Dwight neighborhoods had to travel at least four miles to find a full-service supermarket. Given that these neighborhoods aren't much more than a stone's throw from one of the world's wealthiest educational and medical complexes, Yale University, it was ironic that people in this low-income area had to endure such hardship to secure their daily sustenance. This all changed in July of that year, when the Shaw's supermarket chain opened a store as the anchor tenant in a new shopping plaza called Dwight Place on Whalley Avenue.

The Shaw's was the first supermarket to open in central New Haven in twenty years and now draws a steady stream of customers from the surrounding neighborhoods, Yale University, and other sections of the city. The involvement of the Greater Dwight Development Corporation (GDDC—the same group that the aforementioned Yale MBA students did their research for) made Dwight Place Connecticut's first neighborhood-owned-and-operated shopping center, a fact that instilled immense pride in community residents.

"There was a crying need for a supermarket here," said Paul Ruchinskas, the executive director of the GDDC. "As had other businesses, supermarkets had followed customers to the suburbs." Since the lack of a supermarket had created an aching void in the commercial lives of Greater Dwight's residents, Ruchinskas and his organization's members couldn't have been happier when the store finally opened. "The neighborhoods' reaction was excellent. There is a real sense of ownership. One grandmother told me that walking into the market was almost a spiritual experience."

Before the Shaw's opened its doors, it was estimated that almost all of New Haven's $115 million in annual residential food expenditures were leaving the city for the suburbs. By keeping a much greater portion of that wealth in neighborhoods that are starved for economic activity, the supermarket not only provides people with access to quality food at affordable prices but is also a good-size economic engine. The store employs two hundred people, pays property taxes, and multiplies the economic impact of those food-purchasing dollars as they ripple outward to neighboring businesses. This reinforces the GDDC's mission, which is to improve the economic and social conditions in the neighborhoods as well as to reverse the physical decay they have suffered over the past twenty or so years.

Rather than take a confrontational approach with the supermarket industry, the GDDC decided to work as closely as possible with Shaw's executives and the shopping center developer, McCann Real Equities. "Shaw's was extremely smart," said Ruchinskas. "They haven't come in with the attitude that they don't need any input from the community. They recognize they are moving into a different community. The customers have said, 'This is the first time anyone asked me my opinion.'" Shaw's also agreed to rigid employment targets, which establish percentages of the total staff that will be drawn from the surrounding neighborhoods, the city at large, and then the surrounding towns. It even sponsored a job fair at the GDDC's office for the 200 slots that needed filling. More than 1,800 people showed up.

Perhaps the most elegant part of the deal was the financing. The developer paid a premium price for the 6.8 acres on which the shopping center sits. Because of the high price, the developer then entered into a partnership with the GDDC to expedite the project, which otherwise would not have been financially feasible. The GDDC agreed to purchase the shopping center from the developer for $15 million. To do so, it raised more than $1.3 million through state and federal grants ($1 million coming from the Connecticut Department of Economic and Community Development as a result of the intervention of the Connecticut Food Policy Council) and another $1.5 million in equity from the Retail Initiative, a nonprofit equity fund based in New York City. Yale University helped the project secure a $2.4 million HUD grant from the federal government and provided technical support from the business school. The GDDC borrowed the rest of the money based on the size of the equity stake it now had, the value of the asset—namely, the shopping center—and the projected revenue stream of the tenant leases—the biggest and most stable, of course, being Shaw's. By the time the shopping center opened, the GDDC had also secured a lease with a BankBoston branch and a Hollywood Video store.

Call it shrewd deal making or the shady shenanigans of people who drive around behind tinted windshields, but impoverished neighborhoods got a class A supermarket. The residents gained access to good-paying jobs, and the wider community experienced an economic uplift due to the millions of dollars now flowing through the streets of the neighborhoods. The GDDC, which is now an owner of a significant commercial enterprise, earns income that it can use to develop other neighborhood enterprises. And the Shaw's? Last I heard, the Dwight store was one of its best-performing units in the Northeast. Although it may take a small army of partners and an uncommon amount of business acumen, groups using methods that rely on public financing, philanthropic support, for-profit businesses, and nonprofit organizations can "re-store" America's underserved urban neighborhoods.

THE PHILADELPHIA STORY

Reducing the physical distance between lower-income households and their food sources has increasingly required the intervention of the public sector. Successful efforts to bring full-line supermarkets to underserved communities necessitate a joint partnership between public and private entities, often for-profit and nonprofit organizations. Likewise, bringing people to the food demands an active engagement with the agencies that plan and manage public transportation. In other words, pulling the levers of public policy has become virtually the only effective recourse for those whom the marketplace has failed.

Nowhere have these relationships become more apparent than in Philadelphia, where a nonprofit organization called the Food Trust (formerly known as the Farmers Market Trust) has successfully confronted some of urban America's most severe food access problems. According to a 2001 report from the Food Trust,

> The Greater Philadelphia region has 70 too few supermarkets in low-income neighborhoods. This shortage of supermarkets means that poor residents must travel out of their neighborhoods to purchase food, or shop at more expensive corner and convenience stores with less selection and often poor quality food. The insufficient access . . . reduces the purchasing power of neighborhood residents, and may exacerbate long-term health problems resulting from nutritionally inadequate diets.

Perhaps what is most intriguing about the Food Trust's research is that it found not an insufficiency of sales potential in lower-income areas, but instead an insufficient number of food stores. The marketplace had failed for reasons that have been discussed elsewhere in this book—misperceptions of lower-income communities, racism, more certain opportunities elsewhere, higher development and operating costs—which meant that extra-market efforts were necessary to restore balance to the marketplace.

To address this severe imbalance, the Food Trust recommended that city and state economic development agencies bring together leaders in the supermarket industry to develop a strategy to open more supermarkets. They also prevailed upon government to use public funds "to reduce the risks associated with the development of more supermarkets in lower- and moderate-income communities."

Rarely have the research and recommendations of a nonprofit advocacy group borne such fruit as they have in Philadelphia and throughout Pennsylvania. Under the leadership of the Food Trust, various groups advocated before the Pennsylvania legislature and governor for a state-funded investment pool to be used to increase the number of supermarkets and other grocery stores in underserved communities across the state. Lo and behold, the measure passed, and the Pennsylvania Fresh Food Financing Initiative (FFFI) opened its doors in 2005. According to the Food Trust, "The initiative serves the financing needs of supermarket operators that plan to operate in these underserved communities where infrastructure costs and credit needs cannot be filled solely by conventional financial institutions."

The results have been impressive. The State of Pennsylvania has invested $20 million, which in turn has leveraged $60 million from the Reinvestment Fund, a mid-Atlantic region community investment organization. As of 2006, the FFFI had committed $21.9 million in grants and loans to 22 supermarket projects in Philadelphia, Pittsburgh, Eddystone, and Gettysburg. This unprecedented investment in closing the food gap in both rural and urban communities had created 2,552 new jobs and more than 1.1 million square feet of retail space.

Closing the food gap first requires that the public and private sectors be committed to closing the financial gap. With the cost of new retail development running more than $100 per square foot, the FFFI investment equaled approximately $20 per square foot. Thus, public sources provided 20 percent of the total deal, which roughly equals the conventional investor's assessment of risk. In a simplified explanation of risk assessment, private investors acting rationally and free of misperceptions about urban neighborhoods would say that their financial risk is 20 percent greater in a lower-income area than it is in an affluent area. That is the financial gap, more or less, that government must fill in order to close the food gap.

THE BUS STOPS HERE: LESSONS FROM THE TRANSPORTATION WARS

"Re-storing" America's lower-income communities has been a long, drawn-out conflict that has yet to produce a decisive victor or even a consistent battle plan. In many cases, closing the food gap through the development of high-quality retail food outlets has proven to be an intractable challenge. Sometimes the demographics are too unfavorable, the population density is too low, the necessary skill sets are not available, financing options are sparse, or the supermarket industry will not budge from its cookie-cutter formula for store development. Since people have to eat one way or another, many communities have looked to their public transportation systems for answers. If you can't bring the food to the people, you'd better find a way to bring the people to the food.

Jeanette W. was a twenty-six-year-old single mother of two children, ages three and six, who lived in Hartford's Northend in 1997. I had met her through the WIC Program and some work we were doing with farmers' markets, and she graciously allowed me to accompany her on one of her shopping trips.

To pay her bills, Jeanette worked a part-time school food service job and received about $100 per month in food stamps. Since her youngest child was eligible for WIC, she also received about $30 in vouchers each month to purchase certain high-nutrition food items. Jeanette and her children were typical of many younger households in the predominantly African-American and Hispanic Northend, and when it came to food shopping, she was on the bus like everybody else, heading for the suburban supermarket.

According to the 1990 U.S. Census, almost 40 percent of Hartford households did not own a car. In Jeanette's neighborhood, the number was closer to 70 percent, which lumped her and most of her neighbors into the mildly pejorative category "transit-dependent." Without a car, Jeanette had to do one of the following to reach a large supermarket: plead with the one car-owning neighbor she knew well enough to ask for an occasional ride, take a $10 taxi ride, or use Connecticut Transit, operator of the metropolitan Hartford bus service. Financial circumstances and wanting to keep her neighborly requests to a minimum meant that Jeanette usually chose the bus for her weekly shopping trip.

On most occasions, she could park her younger child with a neighbor, after which she set off on a three-block walk, with her six-year-old son in tow, to the nearest bus stop on North Main Street. There she waited about ten minutes for the next N bus south to Hartford's central business district and transit hub, about a ten-minute ride. Once downtown, Jeanette had to slowly exit the bus to give her son time to negotiate the steep steps down to the curb. She waited another ten to fifteen minutes to catch the K bus west, which took her down Park Street and eventually swung south on New Park Avenue. She was let out almost two blocks from the Stop & Shop, the only full-size supermarket that was physically within Hartford's borders, although its proximity to the nearby interstate and West Hartford made it more accessible to car-owning suburbanites than to car-less city dwellers. From the time she climbed on the K bus until the time she walked through the supermarket's door, another twenty minutes had elapsed. Door to door, it took Jeanette and her son at least one hour of walking, waiting, and riding to reach an affordable, high-quality supermarket. By car, the same trip would have required less than fifteen minutes.

Once inside, Jeanette loaded her now cranky son into a shopping cart and tried to raise his spirits with a sugary juice drink and a small bag of potato chips. Heading down the produce aisle, she selected only a few items —a few green bananas, half a dozen apples, and a head of iceberg lettuce. She told me that her children wouldn't eat things such as broccoli or collard greens but also seemed a little unsure herself as to how to prepare them. In the meat case, she picked up a couple of pounds of hamburger and a small package of pork chops, then she veered into the cereal aisle. Her son perked up immediately at the sight of Tony the Tiger on a box of Frosted Flakes, which she grabbed along with a box of Wheaties (she could use her WIC vouchers for more nutritious cereals such as Wheaties but not for the Frosted Flakes).

At this point, size mattered. The larger items, which were less expensive on a per unit basis, were simply too bulky for her to carry. The same criterion applied to almost every item—family-size hamburger packages, large boxes of detergent, or nearly anything that is cheaper by volume were off-limits to her. Again at her son's urging, Jeanette selected a 99-cent liter bottle of Coke, the heaviest item in her cart. The temperature outside that day was relatively cool, so Jeanette was not too worried about spoilage. But no matter what the temperature, she never purchased ice cream or other frozen food when riding the bus. She also found that most of the leafier greens were usually wilted by the time she arrived home, especially on warmer days.

Jeanette rounded out her purchases with a few more WIC-approved items such as peanut butter, a half gallon of milk, and orange juice. At the checkout counter, she succumbed once more to her son's pleading and reached for a bag of M&M's. She paid with a combination of food stamps, WIC vouchers, and cash. Two separate bags of groceries were carefully packed in double-lined paper sacks placed inside plastic bags with handles. They were heavy, but Jeanette managed to carry them both, one suspended from each arm, while herding her candy-clutching son out the door, across the parking lot, and down to the bus stop.

The trip home was a mirror image of the trip to the store, with similar waiting and riding times, but with the added burden of maneuvering two bags of groceries and an unhappy child. The final three blocks from the Northend bus stop to their apartment was the toughest leg of Jeanette's journey. Between the physical exhaustion and the sugar overload, mother and son had reached their limits of tolerance for each other. A little over fifteen minutes shy of three hours from the time they left their apartment, Jeanette and her son walked through their door, collapsed on their tattered couch, turned on the television, and shared another bag of potato chips and the bottle of Coke.

For the Hartford Food System, stories like Jeanette's were the impetus for a deeper exploration of the role that public transportation played in connecting people to their food supply. By studying the public transit maps of the Greater Hartford Transit District for a few minutes, we quickly realized that there was in fact no connection at all. The bus routes had been designed to move people quickly and efficiently from their homes in the suburbs to their places of employment in downtown Hartford. As long as there were still supermarkets in Hartford, the "spoke and hub" model of single-route vectors linking imaginary compass points at the suburban fringe to the central business district worked perfectly well. But when Hartford's supermarkets moved to the suburbs, city shoppers were forced to use a bus system that had never been designed to take them in that direction. In fact, the sites of many new suburban supermarkets had been deliberately chosen because they were easily accessible by car. After all, didn't everybody own at least one car?

Over the course of two months in 1997, the Hartford Food System and the food policy council that it cofounded, the City of Hartford Advisory Commission on Food Policy, decided to take a closer look at the situation. To learn more, they passed out hundreds of quarters to several college student volunteers and told them to ride the city's buses from the poorest neighborhoods to a variety of suburban locations. Armed with pencils and clipboards, they were instructed to interview bus riders to determine why they were on the bus and what role, if any, buses played in getting them to food stores. The results were startling. More than 32 percent of the people on the city's buses were using them to do all or some of their food shopping. Of this group, 60 percent reported that they experienced difficulty food shopping due to their dependence on buses, a problem that was particularly severe for those who had to use two or more buses to reach their destinations (45 percent of this group). In addition to experiencing the hardship of simply getting to suburban stores, 57 percent of the bus users also had to use small, high-priced neighborhood stores for "fill-in" shopping, an action that often canceled out the price advantage of shopping at the suburban supermarkets.

Additionally, 20 percent of the respondents often used friends with cars to drive them to the supermarket. Sometimes the word *friends* was used loosely and referred to people, aptly nicknamed "vultures," who circled supermarket parking lots during "welfare check" days and offered to take harried, overloaded, and transit-dependent shoppers home for $10 to $15. Some vultures could make up to $300 per day performing this service, especially during the first few days of the month when low-income families, "flush" with food stamps and welfare payments, did most of their shopping. Another option was to take a taxi, which would charge $5 to $10 per ride but was often harder to come by than a vulture.

The report based on these findings recommended that bus routes be redesigned to better serve transit-dependent supermarket shoppers. Since it was also discovered that buses did not have a place for riders to put large items such as grocery bags, it was further recommended that racks be installed for that purpose. And in defiance of all that had been assumed about transportation planning since World War II, the report recommended that the location of supermarkets should be aligned with public transit routes—in other words, food stores should be as bus-friendly as they were car-friendly.

If persuading supermarkets to return to the inner city is difficult, convincing transportation bureaucracies to do something other than build roads is not unlike pushing a passenger bus uphill. Like many government agencies, transportation departments define their mission narrowly, which, in spite of dwindling energy reserves, is to serve America's car culture. Limping along in the breakdown lane is America's public transportation system, which, as anybody who's ever traveled in any other developed country knows, is a sorry case indeed. So imagine suggesting to a transportation official, who is lukewarm to even the concept of "public," that part of his mission should be to help the car-less get to food.

This was the nature of the conversation that food activists in Hartford, and later throughout the state, began with public officials. To their credit, these officials rose to the occasion once they realized that they had a role to play in addressing the community's food needs. Staff members from organizations such as the Hartford Food System and residents of the city's most transit-dependent neighborhoods were invited to sit down with transportation planners to determine what could be done. What commenced was essentially a planning charrette in which both sides—the "bus guys," as I would privately call the transit experts, and the community representatives—explained to each other their respective views of the world. Community folks made it clear how difficult it was to use the bus system to get to health clinics, doctors' offices, food stores, and even jobs that were now located in the suburbs. The bus guys explained how bus routes are developed and paid for.

More than a year later, on August 27, 2000, after studying maps, riding buses together to explore new routes, acquiring such arcane information as the turning ratios of buses, and successfully prying loose additional public funds, Connecticut Transit inaugurated the L-Tower bus route. Together, community groups, policy wonks, and some of the nicest bus guys you'd ever want to meet had carved out a new route that took people directly from one of the city's poorest neighborhoods to a brand-new chain supermarket in the town of Bloomfield. What had been a forty-five-minute bus ride involving at least one transfer was now a direct fifteen-minute trip. To no one's surprise, the L-Tower bus was moving 250 transit-dependent people a day to clinics, jobs, and food and within a few months had exceeded the ridership projections of both the community and the bus guys.

Nearly every urban community in America, and countless rural areas as well, has confronted the failure of the retail food industry to adequately serve its citizens. Even after living with such an egregious failure of the marketplace for thirty years, I'm still incredulous when I think about how nearly an entire industry simply walked away from tens of millions of people without consequence. Unlike the American auto industry, which closed factories because it failed to compete with the Japanese, or the loss of Main Street's small retail shops from rapacious Wal-Mart stores coiled like pythons on the edge of town, the chain supermarkets abandoned vast stretches of this country's landscape simply because they could make more money in more affluent, usually suburban communities.

A thousand schemes have been hatched to "re-store" these jilted communities. The early attempts were often too small, too naive, or too undercapitalized to fill the yawning gap left by supermarket flight. But out of the ashes and assorted misfirings, some successes rose above the rubble, models of public and private partnerships were born, and ideas for replicating the successes emerged. We still have a long way to go before everyone has access to healthy and affordable food, but at last there's reason for hope.

CHAPTER SEVEN

Growing Obese and Diabetic; Going Local and Organic

The corn don't grow so good around the edges,
so this year I ain't planting any edges.
Eighty-year-old Connecticut farmer

When my old farmer friend explained his corn-planting method to me, I of course thought he was pulling my leg. But as time passed, I began to wonder if his remark was a parable spoken by a crusty old fellow known as much for his mischief as for his wisdom. My meditation led me to think that our understanding of communities, people, food, and health are always bringing us up to the edge—we want to know what's beyond those edges, how to push them, master them, or take away their roughness. As individuals, we want to control the edges of our lives that are just out of sight or always in flux. I find myself at times compelled by a fervent hope that I might be healthier, happier, skinnier, or wealthier if I could unravel the mysteries that govern those dark outer limits of my soul. Sometimes we even merge our edges with those of another, which of course eliminates one set of edges but creates a whole set of new ones. In other words, the dance with edges can go on forever and may never satisfy the seeker. They may taunt or tease, occasionally illuminate or suggest, but like the bubble from a child's plastic wand, they always explode when grasped.

Unlike the farmer who decided to avoid the unproductive edges of his fields by simply not tending to them, some people have striven continuously to make their edges flourish by pushing them ever outward. This is the quest that I believe is undertaken by a growing number of Americans who, for the past thirty-five years or more, have been seeking, among other things, better food and healthier, more satisfying lifestyles.

Ironically, their quest is shared by an entirely different group of people whose lives operate under a much less fortunate set of circumstances. Unlike the affluent and well educated, these people's edges are not expanding, glowing, or presenting limitless opportunities. Instead, their edges are atrophying, their choices are narrowing, and their control is eroding. Their edges do not denote a place from which to explore unknown territory or embark on new adventures, but instead form a boundary that can rarely be crossed and a prison wall that cannot be scaled.

Starting in the late 1980s, Hartford's food landscape began the final act of its steady and sickening transformation. As the supermarkets packed up their wares and moved to the suburbs, they left behind a vacuum that was soon filled by the bottom feeders of America's food chain: shiny new fast-food restaurants and gas station mini-marts. As a result, the city's citizens went from being underfed to being overfed in a matter of ten years.

At first glance, given the city's high poverty rates, cheap fast food should be a blessing. If there are no supermarkets within easy reach, people should be grateful for the clean, well-lit places that proffer nicely packaged, brand-name merchandise. But in fact, such establishments thrive in areas of poverty and low education. While they presumably serve a community's immediate need for calories, they actually prey upon those who are weakened by insufficient money, choices, and knowledge. As a result of these factors, Hartford's major food problem shifted from hunger to heart disease, diabetes, and obesity. In light of the soaring rates of diet-related diseases across the nation as well as in Hartford, the high prevalence of unhealthy food outlets became a serious public health issue.

On Saturday, March 31, 2001, more than twenty University of Connecticut dietetics students fanned out across Hartford and two of the city's affluent adjoining suburbs, Wethersfield and West Hartford, to inventory and analyze the contents of two hundred restaurants and small grocery stores. At the behest of the Hartford Food System, they were sleuthing for information about the availability of 100 percent fruit juice, low-fat milk, vegetables (other than tomato, lettuce, French fries, and onion rings), and vegetable entrées. Another part of their mission was to assess the number of high-fat, high-sugar, high-salt snack foods, as well as the overall types of food available and the cleanliness of the establishments. Generally, these young investigators were trying to construct a profile of what constitutes a good selection of healthy food.

The data gathered by the students revealed marked differences in fast-food choices among the three communities. Low-fat milk was available in 50 percent of the sites in the suburban sites but in only 26 percent of the sites in Hartford. This meant, of course, that Hartford residents had fewer opportunities to choose and purchase the generally recognized healthier milk product. Almost twice as many Hartford sites (50 percent) had no vegetables as compared to their suburban counterparts, and one-third of the Hartford restaurants offered only fried vegetable entrées, compared to only 5 percent of the West Hartford and Wethersfield restaurants.

The distribution of fast-food restaurants and other low-quality retail food outlets also was revealing. By mapping the locations of the region's fast-food outlets, the survey found that a very high concentration of them were crouched like predatory cats within easy walking distance of most of Hartford's lower-income residential areas. The proximity of McDonald's, Burger King, and Kentucky Fried Chicken to the region's most impoverished and nutritionally at-risk families was stunning. Positioned as they were along the city's most traveled commercial corridors, they created a virtual ambush for any inner-city resident walking along these thoroughfares.

By contrast, the fast-food outlets in West Hartford and Wethersfield, as they tend to be throughout more affluent and suburbanized areas, were found along commercial strips or in shopping centers that could only be reached conveniently by car. The good news about car-dependent suburbia, where housing developments are spread out and usually located some distance from commercial areas, is that securing a bacon double cheeseburger requires just enough extra effort to make you think twice about whether you really want it. For Hartford's transit-dependent shoppers, who must travel forty-five minutes to reach a decent supermarket, an evening stroll to the corner KFC for a bucket of Colonel Sanders's fried chicken is, sadly, considered one of the few privileges of living in a low-income neighborhood.

One irony associated with this unhealthy food abundance was that it was partially aided and abetted by Hartford city government and other public and private institutions. Poverty created the market, shrewd entrepreneurs took advantage of it, and city hall nurtured the relationship. Shortly after the Hartford Food System completed its healthy food study, the city celebrated the opening of its eleventh Dunkin' Donuts. And *celebrated* is the operative word. Community leaders, representatives of Connecticut Children's Hospital directly across the street, and government officials, including the mayor, showed up to cut the ribbon. The *Hartford Courant* joined the hoopla with an editorial praising the Dunkin' Donuts chain for its "neighborhood sensitive strategy," because the corporation had promised to employ neighborhood residents in its new store. In a city that was struggling to gain any job it could, any business that would provide twenty-five jobs—albeit mostly part-time, minimum-wage jobs with no health benefits—made Hartford's economic development officials salivate.

What wasn't so funny were Hartford's obesity and diet-related disease rates. Based on 1997 data, 44 percent of the city's population was obese or overweight, compared to 24 percent for Connecticut as a whole (these numbers are proportionally much higher today). Hartford's diabetes rate was 8 percent, and its hypertension rate was 28 percent, 75 percent and 50 percent higher, respectively, than the state's overall rates for those illnesses. The rate of diabetes was higher for Hartford's Hispanic population, and the rate of hypertension was higher among the city's African Americans.

Today, 61 percent of Americans are obese or overweight, with the rate of diabetes soaring in every category of race, ethnicity, and age, especially among children. The cost of our rapid weight gain is now estimated by the Institute of Medicine at between $98 billion and $117 billion per year. Obesity is a major risk factor for high blood pressure, heart disease, stroke, diabetes, certain types of cancer, arthritis, and breathing problems. The impact of obesity and its related health complications are so severe that the U.S. surgeon general has expressed concern that this generation of children may be the first in U.S. history to have a shorter life span than their parents.

Lack of exercise, too much television, and an obsession with computer technology have turned Americans into a race of 250-pound weaklings. One of the biggest culprits is fast food. As Eric Schlosser pointed out in his book *Fast Food Nation*, "In 1970, Americans spent about $6 billion on fast food; in 2000, they spent more than $110 billion. Americans now spend more money on fast food than on higher education, personal computers, computer software, or new cars." Why? Fast food is easy, cheap, and readily available, especially in low-income communities. In 2000, the fast-food industry spent $3 billion a year on television advertising, much of it targeted at children. McDonald's and other fast-food restaurants don't rely just on advertising to get young people hooked. As Schlosser also pointed out, McDonald's operates more than 8,000 playgrounds at its U.S. restaurants to lure children onto the premises.

The role of advertising, information/misinformation, and other persuasive mechanisms that disable our rational decision-making processes is apparent in the fast-food industry's growth and the concomitant increase in our waistlines. In lower-income communities, lower education levels and the lack of healthy food choices make households easy targets for fast food's messages, images, and hidden persuaders. Although efforts to restrict junk food advertising directed at children have had some success, there is still not the political will or public budget to compete with the mountains of promotional cash that junk food purveyors have available. Whereas McDonald's spent $500 million on its "We love to see you smile" campaign in 2000, federal spending on the "5 A Day" campaign peaked at only $3 million the same year.

This takes us back to the celebratory moment occasioned by the opening of Hartford's eleventh Dunkin' Donuts. As the Hartford Food System began to publicize the problems of poor access to healthy food and the consequences of the food gap, public officials began to respond. When a proposal later came before the Hartford City Council to sell advertising space, including for Dunkin' Donuts and McDonald's, on the sides of municipally owned vehicles, the cash-strapped city actually gave the proposal serious consideration. This time, however, city officials were much more willing to hear our concerns that promoting unhealthy food products to a city that had the highest diabetes rate in the state was unwise. They agreed to reject the idea.

Such was not the case with Hartford's school superintendent when the Hartford Food Policy Commission approached him with a request to remove all soft drink machines from the public schools. The commission felt that since poor food choices were bringing the city's residents to the brink of a public health crisis, public agencies should do what they could to discourage the consumption of unhealthy food. The superintendent told the group with a straight face that he saw the issue as a matter of free choice and personal responsibility. If students didn't have Coca-Cola machines in the schools, he reasoned, they wouldn't have the opportunity to decide whether to have a Coke or not. Once the group recovered from the shock of hearing such an ill-reasoned argument, it attempted to change his mind. He wouldn't budge. In his opinion, taking soda machines out of the public schools was a denial of the students' rights.

An analysis of the Hartford school system's curricula found that the average city student received about four hours per year of health-related instruction. In addition to diet and nutrition information, this included other issues such as drugs, sex, and personal hygiene. The average child watched television nearly four hours each day, during which they were exposed to copious amounts of junk food advertising. In addition, they were subjected to myriad other advertising sources, including print media, billboards, bus shelters, T-shirts and sweatshirts, and dozens of large, well-lit, and brightly colored soda machines in their schools. How and where were young people to develop the skills they needed to make critical judgments about their food choices when they were assaulted by a well-armed, well-financed junk food industry? It was as if the Hartford public school system was equipping its children with a peashooter to confront a panzer division.

CONNECTICUT BATTLES BIG COLA

When the direct and personal approach fails, as it did with the superintendent of Hartford's public schools, you can always go higher, which advocates did when they took the issue to the Connecticut legislature. When it comes to the fight to prevent obesity, one of the most critical battlegrounds has been in the nation's public schools, and more specifically in the cafeterias and corridors where food is served. Over the past several years, public school districts around the nation have been forced to make financial deals with the likes of Pepsi and Coca-Cola to keep their school cafeterias operating in the black and to give school principals some extra cash to pay for things such as band uniforms. But as one scientific report after another revealed the growing epidemic and health risk of obesity—especially among children—parents, doctors, and nutrition advocates started to fight for the reduction or elimination of unhealthy food from their schools. With a passion you don't often see these days except at a high school basketball game, advocates across the nation started marching on school board meetings and state capitols to demand change.

But Coke and Pepsi are not taking the challenge lying down. Using a page from Big Tobacco's playbook, Big Cola has pulled out all the stops in a desperate fight to halt state legislatures from enacting tough new nutrition standards for public schools. What's at stake for Big Cola? Millions of future consumers in the form of today's children, whose brand loyalty will be embedded in their malleable brains.

Connecticut is one state where citizens started wondering why their schools were being used as junk food feeding tubes for their children. A statewide opinion poll conducted by the University of Connecticut Center for Survey Research and Analysis showed that 70 percent of the state's citizens supported a school soda ban. Under the leadership of state senator Donald E. Williams, president pro tem of the Connecticut Senate, and Lucy Nolan, a fiery advocate for the nonprofit organization End Hunger Connecticut!, both houses of the state legislature passed by overwhelming margins the toughest school nutrition standards of any state in the country in 2005. If the bill was signed by Governor Jodi Rell, no child would be able to buy a sugary beverage in any of the state's public schools. Instead of agonizing over whether to push the Coke or Pepsi button on school vending machines, little Johnny would be able to choose only bottled water, 100 percent fruit juice, or unsweetened milk.

While Johnny may pout for a while, he may very well thank the powers that be many years later when he hasn't joined the ranks of the overweight or obese. And Connecticut and U.S. taxpayers will be grateful when the obesity-related illness portion of Medicaid and Medicare expenditures, which currently stands at $665 million annually in Connecticut alone, begins to decline.

But as Williams and Nolan learned, anyone who goes up against the carbonated commandos of Big Cola faces an uphill battle. Pepsi and Coca-Cola hired the state's most powerful and connected lobbyists to fight the reform efforts. With combined lobbying fees of nearly $150,000, Big Cola's hired guns stalked the halls of Connecticut's state capitol disseminating misinformation and dissembling the issue. Like Iago whispering into the ear of Othello, they told legislatures that the food and beverage menus of local schools are none of legislators business. According to

Nolan, they said that what children eat is up to local school boards and parents to decide, not the state. They even went so far as to suggest that state government would be disempowering children by taking away their opportunity to buy junk food in school.

Big Cola succeeded in obfuscating the issue so well that the Connecticut House of Representatives debated the soda ban for an astounding eight hours. This turned out to be the most time legislators devoted to any single issue in the 2005 session. The disproportionate debate occurred even though the legislature was also considering the nation's first legislatively authorized same-sex civil union bill and an attempt to outlaw state executions (an issue of heightened emotional content since the state's first execution in almost fifty years was scheduled to take place soon). Nolan, the mother of three school-age children, acknowledged that the pressure from soda lobbyists was extreme and unrelenting. "You have to be willing to go up against the state's toughest lobbyist," she said. "Just look at how much money they spent!"

Big Cola won. Governor Rell vetoed the bill. The next year, 2006, it was resurrected, and after some modifications that allowed for the sale of diet soft drinks in secondary schools, it passed the legislature again. This time, Governor Rell signed it.

OREGON AND NEW MEXICO TAKE ON BIG COLA

Such legislative resolve didn't appear to be the case in the 2005 Oregon legislature. Mary Lou Hennrich, director of the Community Health Partnership, led a group of health organizations, medical associations, and academics called the Oregon Nutrition Policy Alliance that put forward legislation to reform the state's school food environment. When the advocates made a few technical missteps, Big Cola's Oregon lobbyists immediately started sowing the seeds of discontent. "They went around to school food service directors and local officials telling them they shouldn't let state government tell them what to do," Hennrich says. Big Cola even brought in one school board official from Eugene who testified against the healthy school food bill. It turned out that he was also the president of a local Pepsi bottling company. Against overwhelming common sense and the health interests of their own children, Oregon legislators defeated the bill.

Joy Johanson, senior policy associate at the Washington, D.C.–based Center for Science in the Public Interest, has seen this pattern time and again. All that state nutrition advocates really want is for schools to offer healthy food. She points out that the food and beverage industry has become quite adept at playing a game of semantics that disorients overworked state legislators and confuses the general public. For instance, the word *ban*, as in *soda ban*, has been framed by Big Cola to mean that Big Brother is treading on our civil liberties.

"The local control message that keeps surfacing throughout this debate," Johanson says, "is not coming from parents or even particularly from local school authorities; it's coming from Coke." She notes that 90 percent of all local school districts don't have a certified nutrition professional on staff, which means that a state-level person qualified to make decisions based on the best scientific information is a boost to local schools, not a hindrance as Big Cola suggests. Of course, no one seems to complain when state boards of education require a minimum number of courses in English and math to earn a high school diploma, or when the federal government demands compliance with its No Child Left Behind regulations. "Healthy school food has become a politically contentious issue when it should be a bipartisan one," Johanson says. "After all, we're talking about our children's health."

But even when Big Cola knows it has to retreat, it finds a way to fight a rear-guard action. After a hard-fought battle in the 2005 New Mexico legislature, the New Mexico Food and Agriculture Policy Council managed to secure legislative consent to establish an expert committee with the authority to promulgate nutrition standards for schools. But there was a cost. The legislature required that the committee members include representatives of the beverage and food industry. In other words, Big Cola convinced the legislature that the fox should join the chicken coop's security guard.

A 20-ounce bottle of Coke contains 16 teaspoons of sugar. Today, according to the American Academy of Pediatrics (AAP), 56 to 85 percent of children consume at least one soda a day in school in spite of the fact that the AAP has declared that soda should not be sold in schools—period. Yet Big Cola and its hired guns spend millions of dollars at all levels of government to retain their perceived right, and presumably that of local schools, to give children their daily sugar fix. It's not that children aren't otherwise bombarded with consumer messages—on television, at the mall, and from their peers. It's just that maybe public schools should offer a sanctuary from life as one big commercial. As Marion Nestle, professor of nutrition at New York University, has said, "There needs to be one place in society where children feel that their needs come first. That's why every aspect of school food matters so much."

There has been a cascade of information about obesity and diet-related diseases over the past decade. The Centers for Disease Control and Prevention (CDC) predicted in 2003 that one in three children born in 2000 will become diabetic unless many more people start eating less and exercising more. Nearly half of African American and Hispanic children, the CDC added, are likely to develop the disease. These projections compare to the slightly more than 6 percent of the U.S. population that has diabetes now. If 45 million to 50 million Americans actually develop diabetes, as these percentages indicate, "there is no way that the medical community could keep up," said Dr. Kevin McKinney, director of the adult clinical endocrinological unit at the University of Texas Medical Center in Galveston. Given that the number of adults diagnosed with diabetes rose by 60 percent between 1991 and 2000 and has doubled for children over the past twenty years, the CDC's projections of future disaster appear to be on target.

PERSONAL VERSUS SOCIETAL RESPONSIBILITY

All the blame for these dire health statistics can't be laid on the doorstep of the fast-food and junk food industries, of course. Many people living in poverty and under difficult social and economic circumstances do escape the clutches of greasy, salty food and for a variety of reasons rise above their surroundings, eat well and live healthy lives. This has led to a debate in some circles about the issue of personal responsibility. Is the responsibility for what one consumes or otherwise does to oneself—whether positive or negative—the person's responsibility or that of society, culture, advertising, the calculating hand of capitalism, or a host of environmental factors over which we have little control?

Though it might be assumed that I favor the societal responsibility side of the debate, I have found myself with one foot planted firmly in each camp. Having worked in low-income communities for thirty-five years, I often feel the kind of frustration I experienced late one night as I stood behind a young, very pregnant mother and her overweight child as she was purchasing cigarettes, Pepsi, candy, and potato chips—and nothing else. I had to wrestle down my urge to rip those things from her hands and admonish her for all the terrible things she was doing to her body, her unborn baby, and her child. (As a community activist colleague said to me once, "The poor screw themselves; the rich screw everyone else.") My liberal tolerance for self-destructive behavior has been tested on many occasions, and I have tried to avoid glib excuses, pat explanations, and the relativistic arguments of the politically correct. After taking a deep breath, I have tried to reflect quietly on encounters with irresponsible behavior to better understand the relationship between an individual's circumstances, his or her personal frailties, and a host of environmental influences. And the more I have done this, the more convinced I have become that, yes, diet-related health problems—one portion of the food gap—can be controlled by personal behavior, but that the social, political, and economic forces bearing down on the most vulnerable are so powerful that it takes an abnormally strong and discerning person to overcome them.

Children eat too much fast food; that's a fact. Do we blame them, their parents, or the seductive scent of greasy burgers wafting across our commercial landscape? Every day, nearly one-third of American children ages four to nineteen eat fast food, according to one study from Children's Hospital Boston. Fast-food consumption has increased an alarming fivefold since 1970. The study estimated that the consumption of fat and sugar associated with such frequent use of fast-food restaurants adds six pounds per child per year and increases the risk of obesity. In the classic struggle between supply and demand, one could argue that the industry is only expanding to keep pace with demand. The Children's Hospital study's findings, however, suggest that the increase in demand is more likely due to the increase in the number of fast-food restaurants and the amount of fast-food marketing.

There is also a relationship between the increase in the consumption of unhealthy food and the decrease in the consumption of healthy food. Obesity is highest among people who eat very few fruits and vegetables, and only one in five Americans is eating the recommended five or more servings of produce each day. Our sense of taste is complex and has been dissected by researchers and journalists alike. Michael Pollan has provided one of the better discussions in *The Botany of Desire*. But surely it doesn't take a panel of Ph.D.'s to determine that a child raised on a steady diet of Big Macs will require an uncommonly creative and patient parent to also hook him or her on broccoli.

While there is still value in debating the causes of obesity and determining the best interventions, the costs and the consequences of doing nothing are readily apparent. A 2005 study by *Health Affairs*, an online journal of health policy and research, found that the cost of obesity-related care to private health insurers rose tenfold between 1987 and 2002. In 2002, that amounted to $36.5 billion and represented almost 12 percent of total health care spending. On average, treating an obese person in 2002 cost $1,244 more than treating a healthy-weight person. In 1987, that gap was only $272. And these are just the costs borne by health insurers and their members. About half of the $98 billion to $117 billion in obesity-related costs every year are paid for by the public in the form of Medicare and Medicaid payments. Poor diet takes a grisly toll on the human body, but it's also taking a financial toll on private insurers and taxpayers.

NEW YORK, NEW YORK: A TALE OF TWO CITIES

"Most of these neighborhoods are not pretty, but the people are like people anywhere else. Spend a little time with them and you'll find out how nice they are." These are the words that Maritza Wellington-Owens uses to describe the New York City neighborhoods where she lives and manages farmers' markets. From East Harlem, her home, to the South Bronx, where she organized her first farmers' market in 1993, Wellington-Owens lives and works at ground zero in the diabetes war. As a woman of Cuban-African ancestry, she's been working for fifteen years to bring fresh, locally grown food to some of this country's sickest communities.

"We have lots of diabetes, hypertension, and poverty in East Harlem," she says. "The food in the neighborhood is for the most part crappy and expensive. You can find lots of carbs and sugar, but you can't find much in the way of whole grains and fresh fruits and vegetables." Wellington-Owens told me that she does most of her food shopping at a Fairway supermarket at 125th Street and 9th Avenue, a major cross-town schlep from her home near 104th Street and 2nd Avenue. She makes the extra effort to shop there because "the $100 I'll spend for food in East Harlem will buy me twice as much on the West Side." This is one of the primary reasons she started the farmers' markets in East Harlem and the South Bronx, two of the poorest neighborhoods in the country. "They are the one place in the community where you can get the freshest and best food."

I can remember the place where Wellington-Owens organized her first farmers' market. More important, I can remember the way I felt when I first saw the site she had in mind—138th Street and Alexander Avenue in the South Bronx—in the winter of 1993. It was bordered on the north side by several twenty-story public housing project towers, whose shabby exteriors and littered grounds were an emblem of urban decay. On the south side of the site ran several blocks of battered commercial storefronts that alternated between boarded up, iron gated, and barely open. Gazing about at what could only be characterized as an inhospitable environment for a farmers' market, I wondered out loud what farmer in his or her right mind would come here. But Wellington-Owens had more faith in her project and community than I did, and with hard work, much cajoling of farmers, and the distribution of at least $10,000 in Farmers' Market Nutrition Program (FMNP) coupons through area clinics, opening day was a success, even though only three farmers showed up.

The connection with the FMNP has been critical to Wellington-Owens's success in organizing six farmers' markets in a variety of low- to moderate-income neighborhoods across New York City. On many occasions, she has been invited into a neighborhood to organize a market by a WIC clinic because the staff knew it would help the young moms and their children eat healthier. Wellington-Owens says that upwards of 50 percent of the sales at markets in lower-income neighborhoods come from FMNP coupons. But it's not just the availability of fresh food that makes a difference to lower-income shoppers. "For those who already understand something about healthy food, the markets make a big difference; for those who are less educated, it takes them a little longer to get it," Wellington-Owens says. "But when they take their first farmers' market tomato home after eating crappy tomatoes their whole life, a light bulb suddenly goes on in their head. The difference is an awakening for them."

For Wellington-Owens and the communities she serves, farmers' markets are a way to get local food to people who need it the most. These are not neighborhoods, however, that can support organic food prices, she explains. "People in these neighborhoods can't spend $4 a pound for tomatoes. That's two gallons of milk! I tell the farmers they can't get $3 a pound here for apples, organic or not. Maybe they can get that at Union Square, but not here. But they learn if they price their products right, they can do well. These are mostly minority neighborhoods where food access is a problem. To me, that's a serious food injustice. I do this work because it's needed and because I see how happy people are at a farmers' market."

In a January 2006 article on diabetes in New York City, the *New York Times* presented a stark contrast between life north of 96th Street and life to the south, along Manhattan's Upper East Side. In East Harlem, whose residents are 56 percent Hispanic and 33 percent African American, the number of people who have diabetes is estimated at 20 percent, and the obesity rate is 31 percent. Just south of 96th Street is the land of the thin, wealthy, and beautiful. There, 84 percent of the residents are white, barely 1 percent of the population has diabetes, and the obesity rate is a slim 7 percent. Who says that wealth doesn't have its privileges?

Of course, the most defining difference between these two large New York City neighborhoods is their poverty rates. More than 38 percent of East Harlem's residents live below the poverty level, compared to just above 6 percent for the Upper East Side. And the food options track those poverty rates very closely. Like the Hartford area healthy food survey discussed earlier in this chapter, a 2004 comparison of healthy food availability in East Harlem and the Upper East Side found that only 18 percent of the food stores in East Harlem carried low-fat, high-fiber food and fresh fruits and vegetables. On the Upper East Side, 58 percent of the stores stocked those items. Food is certainly easy to come by in East Harlem, and with McDonald's offering its Dollar Menu and Kentucky Fried Chicken proclaiming that you can "feed your family for under $4 each," the food is dirt-cheap. The only drawback is that neither the Big Mac wrapper nor the KFC bucket comes with a surgeon general's warning that eating too many of these items is likely to cause a slow, premature death.

The trouble with the food gap and related disparities is that they tend to widen before they narrow. It's a curious market phenomenon that in the United States, where there is so much food and, more important, so much interest these days in ever better and healthier food, that the haves are constantly ascending new heights, while the have-nots continue to discover new depths. Part of the explanation may reside with growing national disparities in income, which can be particularly dramatic in urban areas. Although New York City has been experiencing an economic resurgence for years, the middle class is shrinking, and the gap between rich and poor is widening. New York City's total population by income breaks down as 41 percent high-income, 43 percent low-income, and a mere 16 percent in the middle. If you were a residential developer or a food retailer, which market would give you the bigger return? Naturally, you would gravitate to the high-end market, where demand is indeterminately elastic, and avoid the low-end market, where risk and low profits abound. As the *New York Times* put it, "A two-tiered marketplace can develop: Whole Foods for the upper classes, bodegas for the lower, with no competition from stores courting the middle."

When the marketplace fails to serve everybody with a reasonable degree of equity, health costs in lower-income areas—increasingly borne by the public sector—soar. A 2006 urban Chicago study by Mari Gallagher not only documented the food deserts of lower-income areas but also found a statistically significant link between them and life-shortening health conditions such as diabetes, obesity, cancer, and heart disease. Gallagher's research found that African Americans were the most disadvantaged when it comes to food options. In general, Gallagher determined that residents of predominantly African American neighborhoods had to travel almost twice as far as residents of white and Hispanic neighborhoods to reach a grocery store. The results showed that as access to grocery stores decreased, rates of obesity increased.

As health officials and researchers confirm the gravity of the threat of overeating and unhealthy eating, where does that leave the threat of domestic hunger and food insecurity that has plagued America for the better part of a century? A couple of quick answers may suffice for the moment. Lack of access to healthy and affordable food is a form of food insecurity. If a person can't easily get to sources of nutritious food and/or can't resist the siren song of fast-food and other unhealthy food outlets, food insecurity is a part of his or her life. Additionally, research on hunger, poverty, and obesity suggests the following link: if you don't have enough money to regularly purchase sufficient quantities of food, you will be more inclined to eat high-calorie, filling food to relieve sensations of hunger. Additionally, irregular purchasing power, often a problem in low-income households, leads to binge eating or other irregularities in food consumption, which can contribute to obesity as well.

THE CONSEQUENCES FOR THE POOR

To close the discussion, let's return to Hartford, which has produced excellent community-based research on health and food insecurity issues. In 2003, Dr. Rafael Pérez-Escamilla, a nutrition professor at the University of Connecticut, led a team of researchers in an investigation of diabetes among Hartford's Hispanic population. Diabetes has long been considered a scourge in that community. The team conducted detailed interviews with one hundred women who said they had diabetes and one hundred women who said they did not. The interviews produced some gut-wrenching findings, including the fact that 80 percent of all the women, whether diabetic or not, were obese. This indicated that even among the nondiabetic group, a substantial number of women were highly prone to the disease.

The lack of nutrition knowledge and physical activity among the women was similarly striking. Most of the women in both groups lacked knowledge about how to use the food guide pyramid, and few were able to read a food label. The women in the diabetic group were even less able to read a food label than the women in the nondiabetic group. Most of the women in both groups got very little exercise and reported frequent symptoms of major depression.

The researchers also identified hunger and food insecurity as related factors in the high obesity and diabetes rates of both groups. The women in the diabetic group were twice as likely to report experiencing hunger than those in the nondiabetic group and indicated that they typically had a shortage of fresh, low-carbohydrate food available to them.

Living in neighborhoods that have little or no space for physical activity and recreation, and in the midst of a food desert, they clearly were influenced by the environment and poverty. The toll on such individuals, as well as on their families, friends, and employers, is severe and widespread. These are physically sick people living in sick communities where the hope of lasting, positive change does not burn bright. The marketplace alone is not going to correct this problem. Without a substantial intervention by the public sector and nonprofit groups with a community-wide interest, the people and the communities they live in will remain poorly fed, underserved, and uninformed.

THE POOR GET DIABETES; THE RICH GET LOCAL AND ORGANIC

Here's a cynical thought: maybe it's better to be part of a major national trend, even a negative one, than not to be noticed at all. Of all the problems that the poor face, not participating in some of America's most noteworthy food and health trends is not one of them. As a class, they have been well represented in some of the best-covered food stories of our day—hunger and obesity. But as obesity was becoming such a worrisome issue in the public eye, another food trend that had been slowly maturing for the better part of the twentieth century was rapidly becoming a national penchant—namely, locally and organically produced food.

It's a little hard to pinpoint when and where the organic and local food movements began. Some might credit the Rodales and their gang in Emmaus, Pennsylvania, whose organic farm and garden publications in the 1950s and 1960s gave organic a name and a hazy aura of scientific respectability. Others bow deeply to Berkeley, California, diva Alice Waters for the style and grace she bestowed on something as ordinary as a locally grown tomato in the 1970s and 1980s. While the roles that these and others have played in demonstrating that organic and local are decidedly different from most of the food that is produced by our industrial food system, I tend to credit the likes of George Hall. As a grumpy, eightysomething Connecticut farmer who had been doing nothing but "dunging and digging" his family farm for his whole life, Hall woke up one day to discover that his crops were now worth more because somebody was calling them "organic." Since his farm was quickly being surrounded by upscale suburbanites and farmers' markets were opening up across the state, Hall suddenly started seeing a classier clientele than he had before. "Shit!" he exclaimed one day. "I have no idea who these people are, but that don't mean I don't like their money."

Organic food and organic agriculture had barely climbed out of the bassinet by the 1980s. Anyone who has seen the 1969 movie *Easy Rider* may remember the scene where Peter Fonda and Dennis Hopper tarry for a while at a commune where a drug-soaked band of hippies are attempting to grow food on a piteous piece of ground. Even ten years after that film, the "science" of organic farming hadn't advanced much beyond five-acre farms where hippies flew kites. By 1989, Connecticut's Northeast Organic Farming Association (NOFA) didn't count more than twenty farmers statewide who claimed to be organic producers, and their combined acreage was well under one hundred acres. But a *60 Minutes* broadcast in February of that year changed all that with a report on a chemical spray called daminozide, better known as Alar, which was used on fruit trees and vines, especially apples. The broadcast highlighted a report by the Natural Resources Defense Council (NRDC) that identified Alar, a known human carcinogen, as still being widely used in orchards across the nation. The report and the exposure it received before 40 million television viewers ignited a firestorm of consumer reaction that still burns hot today.

The NRDC report fueled a rise in demand for organic food that has been climbing ever since. It hastened the creation of new organic farms, the expansion of existing ones, and the conversion of conventional producers to organic. Many more producers were motivated to reduce their use of agrichemicals such as fertilizers, pesticides, and herbicides. The demand for organic food couldn't be met by the existing new wave of "hippie farms" or the old "codger farms" of people such as George Hall. Growth was required, but so was commercialization and an accompanying infrastructure that could produce, distribute, and, perhaps most important, regulate organically produced food.

In 1989, Thomas Forster worked for an organic farming promotion organization lobbying Congress to create what would become the Organic Food Production Act in 1990. The passage of the act (which would require twelve years of excruciatingly painful work before finally being implemented) would bring some order to the regulatory chaos that bedeviled the organic movement. "The NRDC report had spiked demand for organic produce that couldn't be met," Forster told me, "and you had twenty-three states all with different organic standards. Now that organic food was moving across state lines, you needed a national standard, without which you'd never be able to meet consumer demand."

The force behind the organic movement was something more fundamental than scientific research and the setting of national standards. As yuppie families began to have children, their food choices were informed by all the same influences that had been swirling around the baby boom generation. And like every parent since time immemorial, they wanted what was best for their children.

What was best for me as a child was something wholesome and homemade by my *Good Housekeeping* mother in the 1950s and 1960s. That ideal had moved to organic by the time I reached my thirties in the 1980s and had shifted yet again to food that was produced locally by the 1990s. And as consumers' food consciousness expands, the market niches narrow, multiply, and subdivide, resulting in an infinite number of food preferences or, in some cases, fetishes. One's food choices may be vegetarian, vegan, organic, grass-fed, free-range, humanely raised, or some combination of these. As to the source of this food, one can be many things as well, ranging from "generally local when it's easy to get" to "obsessively local and will eat nothing else." The debate over what's best can be nightmarishly confusing. As one shopper at New York City's Union Square Greenmarket told the *New York Times*, "Organic was good. Farmers' markets were good. Everything else was not good. Now I don't know how to choose anything. Is it local? Is it organic? Which is better? I don't know."

Shopper anxiety like this has certainly enlivened the chatter at more than one cocktail party or herbal tea klatch, but in low-income circles, it gets little traction. Between getting to a food store where the bananas aren't black and having enough money to buy food, low-income shoppers have little inclination to debate the merits or parse the differences between grass-fed, grass-finished, and free-range. But this doesn't imply that the awareness of organic food is low in low-income communities, nor does it mean that low-income consumers are less likely to buy organic if they have the chance.

LOW-INCOME SHOPPERS SPEAK

One afternoon in 2001, the Hartford Food System asked eight members of Hartford's Clay/Arsenal neighborhood, one of the city's poorest neighborhoods, to discuss their understanding of and interest in local and organic food. The group comprised four Hispanic and four African American residents. Six were women, which included three young moms and three senior citizens. They had agreed to meet to discuss food—what they ate, where they bought it, and what, if anything, they thought of locally and organically produced food. First off, there was an immediate consensus that fresh, inexpensive food—the food they generally preferred—was unavailable in their neighborhood. Everyone agreed that getting to a full-line supermarket was a hassle. As a result, they did their major shopping once or twice a month, and when they shopped, price was their most important consideration.

When asked what the word *organic* meant to them, they answered "real food," "natural," "healthy," and "you know what's in it." While they thought that organic food was preferable to food they described as "processed," "full of chemicals," or "toxic," they said that buying organic food wasn't even an option, because it was simply not available to them. One young woman made a point of saying that she didn't trust the environment where she lived or the food she ingested, noting that the regional landfill, a small mountain that loomed ominously over their neighborhood, adversely affected their health. "Everything gives you cancer these days," she said, adding that she didn't trust the water she drank either. Indeed, the issue of trust permeated the whole discussion. People expressed feelings of distrust toward their environment, community, and food. Conversely, there was an underlying tone of confidence in the safety and healthfulness of food that they could identify as local and organic.

The elderly women in the group had fond remembrances of the rural South or Puerto Rico. For this reason, the idea of locally produced food resonated strongly with them. They spoke animatedly of green beans that "snapped sharp and clear" and fresh garden tomatoes, cilantro, and collard greens, the mention of which brought sweet smiles to their faces. Some of the women had been to Hartford's downtown farmers' market, where they had used FMNP coupons to buy fresh produce. One of the women pointed out that two of the farmers at that market sold organic produce.

The awareness of the benefits of local and organic food, while not clearly understood or accepted by people of all classes and education, including the scientific community, was nevertheless very high among this group. For the elderly, there was the nostalgic association with tastes, places, and times gone by. For those with young children, there was an apprehension that nearly everything associated with their external environment, including food, was a threat. Like parents of all races, classes, and occupations, they wanted what was best for their children, even when they knew that what was best was not available to them. And when it comes to food, there is a fine but resilient thread that stitches together the fears, hopes, and aspirations of everybody who has children.

Market research conducted on behalf of the organic industry confirms my thirty-five years of experience and impressions. The Hartman Group, which has accumulated buckets of data on organic food shoppers and their motivations, has found that Asian Americans and Hispanic Americans are just as likely as white Americans, and in some cases more so, to shop at organic food stores. African Americans shoppers aren't too far behind. Perhaps of no surprise to Hispanic Americans and those familiar with their family-centric culture, 65 percent of what the Hartman Group calls "core organic Hispanic shoppers" have children under five years of age, compared to 30 percent for "core organic white shoppers." As the children get older, the interest in organic diminishes only slightly. Fifty-seven percent of core organic Hispanic shoppers have children up to the age of seventeen, but only 17 percent of core organic white shoppers do. The Hartman Group concludes "that health and a concern for the diet of their children is the pathway to adoption of organics for many consumers."

Organic food consumption now constitutes more than $14 billion in annual consumer sales, a number that has been growing by 20 percent a year since 1990 and is now the fastest-growing segment of the retail food industry. Some people have forecast that sales will reach $30 billion by 2010. The growth is founded on the perception that organic food is more healthful and nutritious because it is both free of pesticides and higher in nutrient content. (It also is felt that organic farming methods are better for the environment.)

The leap to local is a bit harder to understand because the health claims are even more tenuous than they are for organic. But as the Hartman Group has proclaimed with hyperbolic gusto, "There is nothing more organic than marketing local." Food writers have joined the fray as well, anointing "local" as the "new" organic. In the eyes of the Hartman Group, local's cachet is somehow tied to the notion of authenticity and has become caught up in the values and practice of ethical consumption. There is an imaginary bridge of sorts that connects organic and local. The Hartman Group asserts, "There is an underlying belief that organics are also 'good' for the environment . . . and through their purchase support small, family-run farms and businesses." Today's shopper at one of the nation's more than four thousand farmers' markets is likely to assume that the products there are of a higher level of purity, health, and safety than he or she would find almost anywhere else. Many no doubt assume that most products at farmers' markets and all community supported agriculture (CSA) farms are organic as well. There is a stronger inclination among a growing number of shoppers to trust farmers more than supermarkets and the extended global food system.

There is something appealing about local that can't be denied. As the *Economist* put it in December 2006, buying direct (that is, local) "short circuits industrial production and distribution systems in the same way that organics used to." The Hartman Group compares buying food at a farmers' markets to buying a pot at a local potter's shop. Artisanal qualities are evoked, the face of the producer is seen, and the alienation between the consumer and the object of his or her desire is reduced or eliminated. Such shopping experiences seem to be imbued with a modicum of spirituality. If shopping at a regional mall is like descending into the inferno, then buying a product directly from a local farmer or craftsperson is like ascending to paradise. There is, in other words, something transcendent in the passing of an object directly from the hands of the producer to the hands of the buyer.

Had I ever spoken such words to George Hall, his eyeballs would have rolled back and he would have let forth a barrage of cusses that would have stripped the stickers off a raspberry cane. But regardless of the motives that one assigns to farmers or consumers, locally produced food has staked out a wide swath of territory that continues to expand. More than 70 million acres of farmland worldwide are currently certified as organic. The one bright spot in USDA farm statistics is the modest growth in both organic production and the number of farmers who are using direct-marketing channels such as farmers' markets. Consumer surveys in the United States have assessed the preference for locally produced food and consistently find that 70 percent or more of respondents would buy food produced in their area or region if they knew where to find it and could be assured that it was genuinely local. Institutions such as public schools, colleges and universities, and hospitals are clamoring for locally produced food to the point that supply and distribution channels cannot meet demand. At least one health insurance company, Physicians Plus Insurance Corporation of Madison, Wisconsin, is giving premium rebates to customers who are members of local CSA farms. Many better restaurants now place the name of the farm next to each menu item, identifying where the main ingredient(s) come from. (Next time you're in Santa Fe, check out the "Grass-Fed Pecos Valley Ranch Steak" at Harry's Roadhouse.) If chefs are the new rock stars, local farmers are their lead guitarists.

Putting farmers at center stage has been the mission of many local and organic food advocacy organizations. As farmers age (farmers over sixty-five years old today outnumber those under thirty-five by a ratio of 4 to 1) and their overall numbers decline, the question is continually asked, "Who will grow our food?" When America's population was mostly composed of yeoman farmers, the shift from farm to factory, and later from farm to office, was hardly noticed. But scarcity breeds value, and consumers are increasingly placing the dwindling number of farmers on a pedestal.

A 2006 exchange between the author Michael Pollan and John Mackey, CEO of Whole Foods Market, captured a bit of the tension between organic and local. Pollan claimed that Whole Foods, as America's dominant purveyor of organic food, had lost its soul to "Big Organic"—that is, large, corporately owned organic farms such as Cascadian Farms, now controlled by General Mills. By purchasing from such farms, Whole Foods, according to Pollan, had spurned local, independent organic farmers. Although Mackey defended his company's progress in dealing with local farmers, he did concede that Whole Foods could do better. In what amounted to a respectable victory for local food advocates, he agreed to give store managers more discretion in buying directly from local farmers. To that end, Mackey set aside $10 million to be used as small loans to local farmers. While many smaller farmers are skeptical of Whole Foods' intentions, as well as those of other large food buyers, there has been a tilting in favor of location (local) over method (organic).

The farmer's place; his or her use of sustainable, if not organic, production methods; and the size of his or her farm now constitute the holy trinity of selection criteria used by today's most thoughtful food buyers. Interestingly, the use of these criteria, especially the local and size concerns, take us back twenty years to the point where organic (and local) got their first big boost—the development of national organic standards. But missing now, as it ultimately went missing then, is one more criterion that most consumers conveniently forget: social justice.

HOW THE ORGANIC MOVEMENT LOST SOCIAL JUSTICE

As stated earlier, the Organic Food Production Act set the stage for much of the national debate on health and safety issues concerning agriculture and food. The act itself grew out of the necessity to bring order to a hodgepodge of state regulations that would ultimately facilitate the growth of the organic market. To put it simply, the first set of interests was that of the farmer who wanted credit and recognition at the cash register for producing organic food. The second set of interests was that of the consumer who had a need to know what was organic and what was not. As the debate widened, according to Thomas Forster, the concept of the "three pillars" of organic farming was introduced, which maintained that organic farming had to be environmentally sustainable, economically viable, and socially just. The interpretation of "socially just," at least among farmers, was that they had to provide both safe and healthy working and living conditions, as well as adequate pay for their workers. (Forster notes that the treatment of workers was a bigger issue in larger-production states such as California.) But socially just also applied to the ways in which organic farming served the community. And "community" meant a range of things, from ensuring that all consumers, regardless of income, could benefit from organic and that the needs of gardeners and urban producers be considered as well as those of larger full-time growers.

That social justice was put on the table in the first place is a miracle given the innate conservatism of agricultural communities. That it would eventually fall off the table is unfortunate but not surprising. As Katherine Clancy said in "Sustainable Agriculture and Domestic Hunger," "Most farmers' views of the poor seem to reflect a basic conservative ideology with overtones of Calvinism that blames hunger on laziness, or 'a defect in personality or behavior.' " But attitudes like these are not the sole province of traditional Republican-leaning, chemical-using farm organizations. Some of the strongest proponents of local and organic food believe that buying the best, and presumably the healthiest, food, which can easily mean a $4 organic tomato or $15 per pound for grass-fed beef, should be a person's highest priority. There is a kind of moral arrogance that frowns upon a person who chooses to pay $30 a month for cable TV rather than shop regularly at Whole Foods. At its best, the thinking goes like this: "When your values are different from my values, don't come griping to me if you can't afford my product." At its worst, it says: "My priorities and preferences are correct, and yours are not."

Thomas Forster contends that the organic movement was originally committed to social issues, though primarily as they played out in rural development. "As a mostly white movement, we were largely blind to hunger, race, and class issues," he told me. While a few of the smaller farmer advocates, such as New York's Liz Henderson, tried to keep the focus on social justice, the political pragmatists in the organic vanguard decided that they should focus on the unfulfilled market demand for organic food. The consumer appetite for organic was growing, and the government had to intervene to ensure that the market functioned efficiently. That was the political goal. To introduce a host of other social concerns, such as providing an opportunity for everyone to benefit from organic food, might very well have scared away the Republican votes that were needed to create national organic standards.

Forster regrets that social issues failed to make the final cut, but he restates the notion shared by many that the organic movement was never an antipoverty movement. Just as it is not necessarily the responsibility of the supermarket industry to ensure that every low-income neighborhood has a full-line grocery store, sustainable and organic farmers are not obligated to provide low-cost, sustainably produced food to every person. Pricing and business decisions are market driven. Public policy is politically driven. If a person doesn't have money, he or she doesn't participate in the market. If a person doesn't have power, he or she doesn't participate in policymaking, which effectively is the only way to make the market work for everyone. I guess that's the Catch-22.

Interestingly enough, the antihunger advocates mirrored the benign neglect of the organic policy wonks, but from a different perspective. As Clancy said, "The anti-hunger food stamp advocates were not involved in the legislation and lobbying efforts surrounding the passage of the [Organic Foods Production] Act, but were in general not supportive." Their indifference to that debate meant that the farmers and environmentalists working for the passage of the act had no incentive to include low-income issues, particularly with respect to expected higher prices of organic food and their impact on the poor.

Antihunger advocates were no doubt aware of the higher food price projections associated with the growth in the organic market. One estimate in 1989 predicted that a full substitution of organic food in the average American diet would add $1,000 a year to a household's food budget. Though no friend of anything remotely sustainable, USDA secretary Earl Butz had stated in the 1970s that a full-scale adoption of organic farming would leave 50 million Americans hungry. Ever sensitive to the need to protect the Food Stamp Program and the tightly stretched food budgets of low-income households, organizations such as the Food Research and Action Center (FRAC) were not enamored of proposals that might drive up the price of food. Their policy objectives did not include legislation that promoted a healthier environment, and while they were probably not convinced of the health benefits of organic food, their dietary objectives were inclined more toward quantity than quality. At a watershed moment in U.S. food and farm policy, with the 1990 farm bill before Congress and organic farming about to move from adolescence to adulthood, antihunger forces and sustainable farming interests did not even search for common ground. They were like two trains passing in the night.

This lack of connection between the organic movement and the antihunger movement typifies much of what passes for policymaking at the national level. Rarely do Washington food and farm policy gurus arrange bold new marriages between previously unconnected interests. The job of federal policy advocacy organizations and their lobbyists, at least as they see it, is to protect, enhance, and expand their specific programs. Rarely do innovative partnerships between two or more interests gain any traction in the crucible of national policymaking. Not only do antihunger, nutrition, and farm organizations avoid new partnerships and ideas to protect the status quo of their respective interests, but they also may actively resist organizations that are seeking to join with them to promote new ideas. Perhaps startlingly to nonpolicymakers, when partnerships are forged, they are more often in the form of tacit understandings between very odd bedfellows (sometimes referred to as "deals with the devil") that amount to a form of forbearance. In other words, "Don't say anything negative about my program, and I won't say anything negative about yours, even though deep down inside, I truly detest it." For these reasons, innovations do not start in Washington. They usually start at the local and state levels and percolate up. And when it comes to food and farming, local communities generally offer the most fertile ground for planting the seeds of change.

The exponential growth in consumer demand for healthy food that is local and organic is driving a wedge the size of the Grand Canyon between the haves and the have-nots. "In a burst of new interest in food," spouted *Newsweek*'s 2006 food issue, "U.S. chefs and home cooks are grappling with today's mounting concern for health, lower calories and higher nutrition. Americans are demanding—and paying for—the freshest and least chemically treated products available." Whole Foods' John Mackey, whose ever-expanding empire of natural food stores sold $5.6 billion worth of products in 2005, told the *Wall Street Journal*, "The organic-food lifestyle is not a fad . . . It's a value system, a belief system. It's penetrating into the mainstream." The belief system has been taken to new heights by the Slow Food movement. Started in Italy and developing a huge following in the United States, Slow Food has fused local food consciousness to the pursuit of pleasure, to the rousing cheers of its devoted adherents. Although its founder and eloquent spokesperson, Carlo Petrini, has argued for higher prices for food to help farmers, the movement as a whole has been sensitive to charges of elitism. Many of its local chapters, known as *convivia*, have undertaken community projects such as school gardens to demonstrate greater inclusivity.

There is no limit to the number of food products for those who are motivated by taste, environmental concern, animal well-being, political correctness, or simple virtue. Niman Ranch produces a pork to die for, helps pig farmers shake the chemical habit, and costs significantly more than the factory-farmed version. Don't want to spend the "best four years of your life" eating swill from the college cafeteria trough? Select from any of hundreds of colleges and universities that are now featuring "sustainable dining" (some inspired by master chefs such as Alice Waters). These schools may also include food- and wine-related activities designed to educate the student's maturing palate. And when you just can't find anything that satisfies your organic lifestyle where you live, you can always pack up and move. The *New York Times* style page in 2004 featured a number of families who had the financial wherewithal to escape from New York City to the Hudson River valley. Once there, the families "began eating strictly organic foods." One couple said they had moved because the wife was pregnant with their second child and "we decided that the children needed to be in nature." One of the more rewarding aspects of their new location was that most conversations weren't about the latest television shows but about "sustainability and buying locally."

Sounds pretty good. In fact, it just may be the latest version of the American dream. But what about those who can't escape or afford to eat "strictly organic" or for whom "buying local" means the substandard, overpriced food at the neighborhood bodega? How do we begin to make this dream of a healthy and sustainable lifestyle available to everyone?

CHAPTER EIGHT

Community Supported Agriculture

Communities Find the Way

We want extraordinary food for ordinary people.
Will Allen, Vermont farmer

WHAT WASHINGTON LOST while it was making the world safe for organic farmers and the organic lifestyle was social justice. But at the same time that lobbyists were cutting their deals inside the Beltway, a few candles were flickering in the hinterland that lit the way to diversity, inclusivity, and the hope that everybody might share in the promise of health and sustainably produced food. One of these candles, later to become a small bonfire, was community supported agriculture (CSA), which Thomas Forster has noted brought "community" back into the world of organic food. In tandem with the continuing growth of farmers' markets, CSA farms (also known as CSAs) would elevate the primacy of local food and in many cases bring low-income families into the movement.

Community supported agriculture emerged in North America in the late 1980s. It is estimated that today upwards of two thousand CSAs now operate across the United States and Canada. In brief, this is a marketing system that sells each member a subscription, or share, of the season's projected harvest, usually well in advance of the first harvest. Share prices run from $300 to $600 or more depending on the length of the season (some CSAs operate year-round) and the corresponding quantity of food. CSAs tend to be associated with the development of organic and sustainable agriculture because most CSAs are certified organic or have guaranteed their members that they use only organic methods.

Community supported agriculture can be viewed as an extension of the farmers' market concept, which brings together farmers and consumers from small towns to major metropolitan areas. Many farmers find a considerable advantage in operating CSAs because they receive payment for their goods in advance and essentially share the risks of farming—bad weather, pests, low market prices—with their customers. In good years, for example, customers receive more produce, and in poor years they receive less.

Working in partnership with nonprofit organizations and government agencies, some CSAs have been able to use grants and donations to reduce or eliminate the usual share cost for low-income families. Other CSAs have come up with a market basket of creative subsidy mechanisms, from offering "working shares" to lower-income members, to holding fundraising events at their farms, to simply asking their higher-income members to contribute a little extra money. These approaches and a respectable tendency toward inclusivity have reduced the perception that CSAs, as well as organic and locally produced food, are the special province of a moneyed elite.

CSAs have contributed mightily to consumers' understanding of where and how food is produced, which of course reinforces the identity of locally produced food. The notion of "community" supported agriculture certainly brings the producer and the consumer closer together, but it also makes it clear that the food and the farm are associated with a particular place, not just the amorphous global food system, whose places and producers are nameless and faceless.

CONNECTICUT CSAS WORK FOR EVERYONE

The fourteen-year-old Holcomb Farm CSA is located in Granby, Connecticut, about thirty minutes from downtown Hartford. Stop by on a September morning, and you'll see overflowing crates of just-picked organic produce stacked high and deep in the farm's distribution room. The wooden shelves in the farm's converted tobacco barn strain under the weight of lush collard greens, the season's first winter squash, and at least six varieties of tomatoes, some so large you could barely balance one in your hand. The farm's manager, Sam Hammer, whose dramatic name is matched only by his imposing six-foot four-inch frame, rolls open the barn doors to a small crowd of waiting CSA shareholders. Within seconds they are milling about the barn, murmuring appreciatively as they eye, caress, and eventually place their produce items reverently into an assortment of cloth and recycled bags.

Like CSAs everywhere, this scene deviates sharply from those in full-size supermarkets, where the produce is barely touched, never sniffed, and thoughtlessly tossed into a shopping cart. At Holcomb Farm, smiling, fresh-faced interns stand by proudly to tell members about the produce they picked only hours before. At the supermarket, a desultory clerk tosses the produce onto shelves and cannot tell customers anything about its place of origin or method of production.

Outside the pickup barn at Holcomb, in an adjoining meadow that serves as the farm's temporary parking lot, the shareholders' Volvos, BMWs, and Outbacks sit side by side with several dilapidated vans that bear names such as Community Partners in Action, City of Hartford Housing Authority, and the Hispanic Health Council. They represent the Hartford community organizations that work with the city's low-income households, and like other members of the CSA, they are here to pick up their share of the produce. The diversity of motor vehicles makes a statement that everybody, regardless of race, residency, or income, is entitled to a share of the best and healthiest food the community has to offer.

The right to food—including good local and organic food—is a principle to which the Hartford Food System holds firm. During the twenty-five years that I ran that organization, I learned that food is not just about putting enough calories in a person's stomach to keep him or her alive. It has a much more complicated relationship to the culture, health, and quality of life of the community. As one middle-aged African-American woman told me at one of the CSA's potluck suppers, the CSA gave her extra-high-quality produce that she couldn't get anywhere else: "Try finding organic collards in a food store in Upper Albany [the Hartford neighborhood where she lived]. And I can't afford to buy them at a natural food store in the suburbs."

Many farmers also recognize that food is not just a commodity that they exchange with consumers for cash. Tony Norris runs a four-acre, urban organic farm in the heart of a lower-income neighborhood in New Britain, Connecticut, home of the Stanley Tool Corporation. He has studied the food preferences of the different ethnic groups that live near his farm so that he can grow what they want. "What we've learned," he says, "is that all the groups we serve—Hispanic, Polish, African-American—place special value on certain foods that are central to their ethnic traditions and, because of their own rural and agricultural roots, place special value on food bought directly from a farm."

This is no small point. Many CSAs and farmers who sell at farmers' markets are responding only to the food preferences of an educated, white clientele. To be inclusive of a more racially and ethnically diverse customer base, farmers, most of whom are white, have to learn how to grow crops preferred by nonwhite customers. Price remains an issue, but if customers don't want what you grow, they won't even bother to show up.

Norris is an excellent producer and an astute businessman, who has learned how to market to a very diverse clientele. But without support from the public or charitable sectors—in his case from the Farmers' Market Nutrition Program (FMNP)—his customers wouldn't be able to pay the price that he needs to charge to support his farm business.

Food, of course, tells many tales about people—their cultures, traditions, lifestyle preferences, homelands, and personal health, to name a few. Like the clothes we wear and the cars we drive, food is an expression of who we are and what we value. For the Volvo-owning CSA members who return home to some of the most affluent towns in Connecticut—the state with the nation's highest per capita income—their food, like their cars, is a deliberate and well-considered choice. They belong to the CSA because they believe it will enhance the health and safety of their families, protect the environment, and support local agriculture. In the late fall, after the CSA's last turnip and Brussels sprouts have been doled out, these thoughtful and earnest folks will resume shopping at Wild Oats or Whole Foods. There they will pay considerably more than they would at a conventional supermarket for the same assurance of health and safety that they derive from the CSA's organically grown food.

For Hartford's lower-income residents, food tells a different tale. It is one whose cultural roots grow deep into the soil of their homelands in the American South, Puerto Rico, and, more recently, Bosnia, Kosovo, and Sudan. The link between food and land are not distant third-generation memories, as they tend to be for the more affluent CSA members. For the poor, the agrarian past has sometimes left a fresh and many times painful impression. While their tales may contain sweet remembrances of a pastoral life, they also may include stories of food scarcity, malnutrition, and hunger. And their new urban homes, even for those now two or three generations removed from rural life, have not necessarily provided sanctuary. In many cases, rural poverty has simply been exchanged for urban poverty.

It was for these reasons that the Hartford Food System organized the Holcomb Farm CSA in 1993. The organization was inspired by Robyn Van En's pioneering Indian Line Farm CSA in western Massachusetts, which she founded in the late 1980s. She modeled the current CSA concept on Japanese farmer-consumer collaborations, which made knowing the person who produced your food a prerequisite. As a reaction to the growing concern about agrichemicals in the food chain, the CSA method of farming and direct marketing mushroomed after Van En's early efforts. If you knew the farmer and had confidence in his or her farming methods, why not enter into a direct relationship with him or her that would benefit both parties? This arrangement suited those who could afford to pay the several-hundred-dollar up-front share price, often due months before the first spring greens were up. But it didn't fit the needs of people who relied on food stamps and food banks for their daily meals. As the Hartford Food System had done in the 1980s when it organized farmers' markets in low-income areas and later started Connecticut's farmers' market nutrition program for low-income families, it saw the promise of the CSA movement to engage the city's poor. All that was needed was a farm, a farmer, and the resources to pay for it.

With the good fortune that sometimes comes to those whose faith exceeds their skills and wisdom, all three of the necessary ingredients were found. Starting in 1993 with 5 acres of land in Granby, 35 family shareholders, and 5 Hartford nonprofit community organizations, the Holcomb Farm CSA has grown to 25 acres, 300 family shareholders, and 11 community organizations. Many stories speak to the development of the CSA, but none is more illustrative than how the municipality of Granby decided to allow Holcomb Farm, which had been bequeathed to the town by its original owner, Tudor Holcomb, to be used by the Hartford Food System. At a town meeting where the fate of the farm was to be determined, a local citizen rose from the crowd of more than two hundred people. He said that first and foremost, Holcomb Farm should be farmed (alternative proposals had included a golf course and a subdivision of McMansions), and it should be used to help "feed hungry people in Hartford." Both purposes won unanimous public support that night as residents overwhelmingly dismissed their reservations about working with a city that was becoming better known for crime and poverty than for any other attribute.

Beginning with these publicly stated values—that the land should be productive and that it should serve all—a network of relationships emerged that would include 25 towns and cities, hundreds of businesses and organizations, and thousands of individuals. In a typical year, the CSA now produces 150,000 pounds of 100 varieties of organic fruits, vegetables, herbs, and flowers. Between 30 and 40 percent of this produce is distributed among 1,200 low-income Hartford residents through 11 community organizations. Operating as a profit/nonprofit hybrid, the CSA sells market rate shares at $475 per year to its family shareholders and deeply discounted shares to its organizational members. While the Hartford Food System raises $50,000 a year to ensure the inclusion of low-income people in the CSA, middle- and upper-income families make an important contribution to the farm's operating costs and maintain the socioeconomic balance that is at the heart of the CSA's mission.

Community goodwill and charitable dollars may keep the farm's tractors running, but it is agricultural skill and experience that make the crops grow. That's where farm manager Sam Hammer and his team of college-age apprentices come in. Farming know-how is a critical component of any farm operation, but in the case of an organic farm that must be environmentally sustainable—that is, respects the limits of nature—the farmers must have a wider command of biology and chemistry to add to their usual store of farm wisdom. If the quality of food is not high, the members won't return; and if the members don't return, the bills won't get paid. Add to good farming and business management practices the task of meeting social goals such as growing high-quality organic food for low-income families, and you have the triple bottom line that constituted the original underpinnings of organic agriculture.

Fortunately, Hammer juggles the competing demands with aplomb. He patiently educates the members about the carrying capacity of the land, the limitations of Connecticut's growing conditions, and the farm's broader community mission. He admits that not all members are moved by these concerns. Some don't understand why they can't get strawberries all season long, and others question why the farm should subsidize the participation of low-income people. He also knows that the problems of hunger and unhealthy eating are immense. "The CSA is only one small piece of solving the food problems of low-income people," Hammer readily confides. "There will always be a need for subsidized food." But he's proud of the fact that top-notch organic produce is going to some of the neediest people in the region. "Organic produce doesn't have to be as elitist as it has become," he says.

A CSA IN NYC

The experience of the Holcomb Farm CSA raises important questions, both practical and philosophical, as to how society can meet a triple bottom line—a food system that's good for farmers, the environment, and *all* consumers. Maybe the biggest question of all is who will pay for that kind of food system. So far, nonprofit organizations and their donors are the ones footing the bill, even in areas where the cost is the highest. Take the Bedford-Stuyvesant section of Brooklyn, New York. In the middle of what can only be described as a model food desert, you'll find St. John's Bread and Life program, which serves Brooklyn's poor with a soup kitchen, a food pantry, a health clinic, employment counseling, and legal services. The program's motto, "Caring for the Hungry," is emblazoned on its sign. According to executive director Larry Gile, "Food is at the core of what we do and the way that we reach people." The program's statistics attest to the need for food: its food pantry served 87,000 meals in 2002 and 133,000 meals in 2003, an astounding one-year increase of 53 percent.

On a hot July day in 2004, Yemi Oyename, Bread and Life's food service director, was overseeing the organization's newest program—a weekly CSA farm produce delivery. Her head wrapped in a stylish blue and silver scarf, the Nigerian-born nutritionist was kibitzing with upstate farmer Ted Blomgren. He was unloading fifty cases of produce directly opposite Bread and Life's main entrance after driving three hours from his Troy, New York, farm. Folks from five nearby neighborhood organizations were waiting to pick up their share of his lettuce, sweet corn, golden beets and squash. Their double-parked vehicles ran the gamut from a 1969 Cadillac to the distinctive passenger vans of storefront churches with names such as the Savior's Shining Light Chapel. An obese woman maneuvered her electric wheelchair through a crowd of people waiting for lunch at Bread and Life's soup kitchen. She nearly ran over an elderly man snoozing against a railing in the noonday sun.

Never breaking a sweat, Oyename coolly orchestrated a scene that threatened to descend into chaos at any moment. "We're doing this because we need a total reeducation of how people cook and eat around here," she said, pointing to Blomgren's fresh organic produce and referring to the nutrition education activities that she had designed.

In a community that has few good food stores and where obesity and diabetes are well above the national average, Bread and Life has its hands full. Undaunted, Oyename employs a number of clever devices to excite nutritionally vulnerable people to the wonders of fresh produce. She uses more common names to refer to some of the vegetables—such as Chinese cabbage instead of bok choy—runs support groups for her clients to familiarize them with the CSA's produce, and features a new vegetable each month (July was squash month) in cooking demonstrations. The previous winter, Blomgren had sat down with Oyename and Larry Gile to select seeds for crops that would appeal as much as possible to the food preferences of Bread and Life's clients. Because of this early consultation, Blomgren added extra rows of collard and mustard greens, planted a little more cilantro, and cut back on the arugula. By making good food affordable and accessible, and by promoting its consumption through food education activities, Bread and Life hopes to reverse the unhealthy diets brought on by poverty, food insecurity, and poor food access.

The genesis for this CSA program was the forward-thinking nonprofit organization Just Food. Based in New York City, the group believes that everyone should have access to healthy and affordable food, especially food that is produced locally and sustainably. Led over the years by three dynamic women—founder Kathy Lawrence, Ruth Katz, and now Jacquie Berger—Just Food has built a network of farmers and community organizations that has accomplished two objectives: (1) made it practical and profitable for farmers to bring their food to New York City neighborhoods and (2) made it desirable and affordable for community groups to buy that food. In addition to using a variety of old-fashioned organizing and outreach methods, Just Food has secured sufficient financial support to subsidize the participation of low-income people and community organizations such as Bread and Life. Part of its support comes from higher-income members (as at the Holcomb Farm CSA), which promotes diversity and distributes the financial burden. The results have been impressive: as of 2006, Just Food had organized more than 40 CSAs serving 10,000 participants in all five New York City boroughs. It also supports community farms, farmers' markets, and food education programs.

CATHOLIC CHARITIES AND PETER'S RETREAT: SERVING THE HARD-TO-SERVE

With the active intervention of an intermediary such as the Hartford Food System, the best food can get to the most nutritionally vulnerable people. Among the Holcomb Farm CSA's members are two Hartford organizations that serve two segments of the community for which diet and health have special exigencies. The first is Catholic Charities' Migration and Refugee Services, part of the Catholic Archdiocese of Hartford, which assists several hundred refugees a year who have come to Connecticut from places such as Bosnia, Somalia, Sudan, and Liberia. They arrive malnourished and destitute from United Nations refugee camps, which have been their "homes" for as long as ten years. The second organization, Peter's Retreat, is a twenty-six-bed congregate-living facility for homeless men and women with HIV/AIDS, many of whom also are making the transition from prison and/or substance abuse. The facility provides a full range of care, including two meals a day, seven days a week. From the time that Peter's Retreat opened in 1994 to the late 1990s, almost half of its residents died each year, a tragic number that has been reduced substantially in recent years as a result of advances in medications.

Sister Dorothy Strelchun, who directs the Catholic Charities program, credits the Holcomb Farm CSA with giving her clients a way to gradually adapt to American culture and food. While the fruits and vegetables are important contributions to the refugees' household food options, the local food also reinforces traditional connections between land and agriculture. Since many of the Somalis and Bosnians were gardeners or farmers in their homelands, their visits to the farm help them make the connection between the food and the soil where it is grown. The outings to the farm also are reminiscent of the small-group experience of going to market in their native countries, where "market days" are an integral part of the social life. "We have to study every refugee group we work with," Sister Dorothy told me, "because we know how important food is to each culture, particularly how it brings people together."

When Henry David Thoreau wrote, "The fruits of New England . . . educate us and fit us to live here," he probably didn't anticipate the repeated waves of new immigrants crashing onto America's shores. But Catholic Charities finds itself following his advice as it tries to "educate" the palates of the region's newest residents. As people who have lived with scarcity for so long, they have a tendency to overcompensate when surrounded with the unfamiliar abundance that Americans take for granted. For instance, Sister Dorothy gives a starter package of staple food items, including a five-pound bag of sugar, to each arriving family. In several cases, mothers have had to hide these bags from their children after they found the children scooping handfuls of sugar directly into their mouths. The CSA's strawberries, raspberries, sweet corn, and apples offer an opportunity to teach these new families healthier ways of satisfying their natural craving for sweetness, while also acquainting them with the foods of New England.

At Peter's Retreat, the CSA shares of collards, tomatoes, and peppers don't go to individual households; they go instead to the facility's communal kitchen, which feeds residents whose lives at society's margins may very well qualify them as refugees in their own country. According to Ron Krom, the organization's executive director, "There are many illnesses and diseases related to HIV that create complications, like diabetes, that can be treated by diet." He says that the Holcomb Farm's produce gives the chef, Janet Candela, more menu options, and as the residents are introduced to the farm and its food, they take a greater interest in their diets. Krom doesn't think that the residents attach any great significance to the food's organic attributes, but he does believe that they make a stronger association between the local farm, its land, and the food they eat. If anything, the food's organic quality is just icing on the cake.

Although Candela is a vegetarian, she readily admits that all of her "customers" are meat-and-potatoes people. "These are folks who are so far off the mark from health," she told me. "Their days consist of sleeping, taking meds, and eating, so I have to feed them what satisfies them." Rather than trying to force vegetables down their throats, Candela arranges weekly field trips to the CSA for six or so residents at a time. At first these trips were just a way to get out of the city and experience a place that was green, but before too long, the residents began to refer to the half-hour ride to the farm as "a trip to paradise." Once there, they pick strawberries, peas, and tomatoes; hang out with the resident dog; or hold the farm's chickens. One day they picked hundreds of stems of cosmos, calendulas, and asters for a staff member's wedding. "I never thought I'd see these guys cut and arrange flowers," Candela said.

Before too long, Candela didn't have to push vegetables down their throats; the residents were eating them of their own accord. Walking on the land that produced their food and helping in small ways with the harvest have inspired them to modify their diets. "The farm became central to my cooking—an organic experience, no pun intended," Candela said. "It had a huge impact on the residents' diets." She noted that the residents have developed a spirit of "living off the land" and being tied, perhaps symbolically, to the farm's seasonal cycles. They ask her in the afternoon, for instance, what vegetables she'll be using for supper. Candela attributes the overall increase in their consumption of produce to the elemental connection between the object and human touch. "If you put a pumpkin seed in the ground and grow and pick it yourself," she explained, "you'll like it much more than the one you buy at the supermarket."

Alani Willett is a Peter's Retreat resident who tested positive for HIV while living in Hawaii. In 1997, after moving to Connecticut, he was diagnosed with an "AIDS-defining brain infection" and given two months to live. He has defied that prediction and stays active by taking college courses and tutoring at an inner-city elementary school. Nevertheless, the disease has taken away a quarter of his brain function, which has reduced his tolerance for stress and at times makes him disoriented. Willett became a regular rider with Candela on their trips to the farm and shared with me the following passage from a journal he keeps.

> It has been raining steadily for two days. This afternoon it is still raining, but gently now. The drops slide languidly off my straw hat as I squish down the row. The raspberries are the size of my thumb-tip. From each fruit hangs a large, lazy drop of life-giving liquid. A gentle tug, and the berries drop into my palm. Soon the pressboard carton in my left hand is saturated and threatens to bombard the earth with my harvest. Reluctant to leave its parent, an adolescent berry drags along an inch of stalk—better eat it! I eagerly strip it with my teeth and savor the tart sweetness.

Candela reflected on the impact that she believes the CSA has on her clients, who "are poor and isolated. They have burned themselves, and they have been burned by society. For them, the farm eliminates the class differences. They mingle freely there with many kinds of people and don't feel out of place. It's a funny thing to say about tough guys from the street, but they feel safe there. And maybe there's something about this environment that promotes feelings of safety, acceptance, and health. The rows of green plants, no pesticides, the clean soil, the clear babbling brook, caring people, and the land itself practically create a craving for healthy food and even a healthy lifestyle."

Should experiences like these be reserved only for select classes of people, especially when the potential for healing is so enormous? Granby's residents have opened their community and made their land available to all the members of the Holcomb Farm CSA. It is a model that other communities would do well to follow.

CHAPTER NINE

Public Policy

Food for the People

It is not only our duty to make the right known,
but to make it prevalent.
Edmund Burke

Everyone takes sides in social change if it is profound enough.
Wallace Stevens

GENERAL CHARLES DE GAULLE ONCE SAID, "Politics are too serious a matter to be left to the politicians." While I wouldn't want to leave politics to the generals, I take this statement to mean that citizens should never forfeit the process of making public policy to their elected officials. Once we, the people, have elected policymakers, we need to hold them accountable for their actions, whether in matters of war, health, or hunger. When it comes to hunger, food insecurity, nutrition, or agriculture, I can say with categorical certainty that not a single significant social or economic gain has been made in the past fifty years without the instigation and participation of an active and vociferous body of citizens. Innovation and change do not occur spontaneously in major centers of political power without the incendiary spark of the populists' voice.

Democracy works best when it's closest to the people. That is why we can expect city hall to act faster than the state capitol, which in turn tends to respond to its people before Washington, D.C. The farther away the decision makers are from those whose lives are affected by their decisions, the slower will be the change that occurs. There is arguably a value in lengthening the time between the demand for change and the institutionalization of that change through means such as law, regulation, or budgetary action. Our body of law necessarily lags behind societal changes and needs, which is as it should be. The institutionalization of change is serious business and deserves the careful consideration of cool and collected heads. But as we look back over the United States' efforts to end hunger and provide equitable access to healthy and affordable food, we see that the food gap has widened or narrowed in proportion to the amount of citizen pressure applied. An informed and activated citizenry, one that speaks for the grass roots first and foremost, is necessary to secure lasting change in this country.

I would hasten to add a corollary to the good general's admonition, which is that policy and politics are also too serious a matter to be left to special-interest groups and their professional advocacy organizations. For too long, organizations that represent narrowly defined business and public interest concerns—whether it be arms manufacturers, zucchini growers, or food stamp advocates—have dominated the national policymaking scene. Their emergence over the past thirty years, especially in the world of hunger, nutrition, and agriculture, have been necessitated both by the size of what's at stake and by the complexity of the federal policymaking apparatus. When billions of dollars can be gained or lost, when millions of lives can be affected for good or ill, and when an unnatural wizardry is required to navigate the labyrinth of Washington, it is no surprise that a professional cadre of advocacy organizations and lobbyists has ascended to such prominence.

But the sheer dominance of well-oiled national advocacy machines has had a chilling effect on those working at the local and state levels for change. Ideas and collaborations that have been nurtured in communities or at the grass roots rarely get a fair hearing inside the Beltway. Robust debates between groups that one would assume have common interests rarely take place. Joel Berg, one of the nation's fiercest antihunger advocates, recently proposed the creation of a single national food program that would effectively create one application for food stamps, WIC, and child nutrition programs. His proposal also would have increased the eligibility for this program to anyone who earned up to 185 percent of the poverty level. While this would have simplified the administration of programs that now have their own applications, procedures, and bureaucracies, it also would have increased the amount of benefits by billions of dollars, since the eligibility maximum for the Food Stamp Program is now 130 percent of the poverty level. According to Berg, "The Food Research and Action Center [FRAC] was afraid that right-wing politicians would use this as an opportunity to gut these nutrition programs."

In written testimony before the House Appropriations Subcommittee on Agriculture in the early 1990s, the Center on Budget and Policy Priorities (CBPP) opposed the addition of a few million dollars to the budding Farmers' Market Nutrition Program (FMNP). The CBPP was afraid that such a request would take money away from the multibillion-dollar WIC Program. Organizations such as FRAC and the CBPP have emerged over the years as pit bull–like defenders of one or two single federal food programs. Each of these programs has its own federal bureaucracy, professional associations, and even for-profit interest groups (for instance, the banking industry for Electronic Benefit Transfer, or EBT, and infant formula companies for WIC). An attempt to reform these programs or introduce even the most modest innovation is vigorously opposed by a small but formidable phalanx of vested interest groups. Change in these programs, when it occurs at all, doesn't amount to much more than tinkering at the margins.

Certain hunger and social welfare groups are directed by a kind of self-anointed directorate whose wisdom on all things Washington is viewed as unassailable. These groups are rarely accountable to any significant constituent base or have members who merely rubber-stamp their actions. Simply by dint of their longevity in Washington, national antihunger, nutrition, and agricultural leaders have become so indoctrinated in the internal logic of Congress, with its gentlemanly pretext and ancient rules of conduct, that those who might have an alternative proposal or method are screened out. One can imagine an average five-foot ten-inch college basketball player brought in to play with an NBA team. No one would ever pass him the ball or give him a shot at the basket.

Such an environment has not provided for the kind of democratic give-and-take that one would expect from those who are fighting for the public interest. If a well-respected Washington think tank were to write a comprehensive plan for unifying and reforming the fifteen separate federal food assistance programs, and in so doing recommended a single funding level that was high enough to effectively eliminate food insecurity, I would wager a bet that it would be vigorously opposed by Washington's various antihunger institutions. Because of the primacy of public policy in ending hunger and making healthy food available and affordable to all, and because of the failure of both the federal government and national advocacy groups to achieve those goals, nonprofit and other public interest organizations have turned to their own backyards. While nothing will ever supplant the massive infusion of federal dollars that pour out across the nation every year to meet people's nutritional needs, it has been left to the innovative spirit and radical vision of local actors to do what national leaders have failed to do.

SMALL STEPS LEAD TO BIG CHANGES

It was the winter of 1987, and I was sitting in a cushy leather chair in the office of the Connecticut commissioner of agriculture. I was there because I had been increasingly concerned that the farmers' markets we had established were not meeting their stated objective of providing local produce to Hartford's low-income families. In addition to the markets we had started in and around Hartford, several others had caught on in other Connecticut cities such as New Haven, Stamford, and New London.

Things weren't going according to plan, however. In one sense, the classic market economy was performing well. We had reintroduced farm fresh produce into people's lives, and the more affluent were snapping it up as fast they could. But the marketplace was not serving those with limited means and transportation, and as usual they were left on the sidelines. I had heard about an experiment that had taken place with Massachusetts farmers' markets during 1986. Massachusetts commissioner of agriculture Gus Schumacher and his associate Hugh Joseph had used several thousand dollars of state funding to give coupons to participants of the federally funded WIC Program. These coupons could be redeemed only at approved farmers' markets. The experiment had gone exceptionally well. Almost 70 percent of the coupons had been redeemed, a user survey found that the WIC participants had started eating more produce, and the farmers were very pleased with the additional income. Why couldn't we do something like this in Connecticut?

Up to this point, I had developed and managed programs all over

Hartford. My work was in neighborhoods, with private organizations, and increasingly with corporate funders such as Hartford's insurance companies. I had not gone out of my way to avoid government agencies—I spent time in city hall and in state government buildings when necessary—but I generally found these systems cold, intimidating, and hopelessly political. In the case of agriculture, the one or two commissioners I had come in contact with had had about as much shine as a rusty dairy bucket. Their idea of innovation was that today we'll milk the black-and-white cows first, and tomorrow we'll milk the white-and-black cows first. More important, they did not see their jobs as being about the work of reinvigorating the state's moribund farming sector. Instead, they acted as if they were presiding over its funeral. And it was into this somber tomb of a state office buildings that I had come, aching with trepidation and hesitancy, to ask for money to do in Connecticut what had been done in Massachusetts. You could say that this was my first serious foray into the world of public policy.

I was pleasantly surprised by the outcome and more than a bit flabbergasted by the process. The newly appointed commissioner, Ken Anderson, was a former Connecticut Farm Bureau lobbyist and journalist not much older than I. He had a curious twinkle in his eye (a characteristic of most good politicians, I would later discover) that made me believe that he genuinely cared about my thoughts. Sitting opposite him across a giant mahogany conference table, I gave him my halting pitch, which concluded not so gracefully with my request for $20,000 to start a Connecticut farmers' market nutrition program.

Lo and behold, Commissioner Anderson thought it was a great idea. I let out an audible sigh of relief. "But," he said, "I recognize that farmers will benefit from this program as much as WIC participants. So I'm willing to put up $10,000 if the commissioner of health is willing to put up $10,000." While it seemed reasonable that the Department of Health, which ran the state's WIC Program, should share in the cost, I wasn't sure where Anderson's challenge left me. Before I had time to think about my next step, the commissioner stood up, grabbed his coat, and said, "Let's go see him." I couldn't believe that he was proposing to walk in on the commissioner of health unannounced and ask for $10,000, but that's exactly what we did.

Once we had entered the health commissioner's office, Anderson told me to relate my "interesting proposition," which I did only slightly more fluently than I had thirty minutes before. The health commissioner immediately agreed to match Anderson's $10,000, and the new program was up and running. I left the building with a lightness in my step and a song in my heart. Gee, I thought, don't we live in a great country!

The Hartford Food System piloted the farmers' market nutrition program in 1987 in collaboration with those two state agencies, several farmers' markets, dozens of farmers, several WIC clinics, and an assortment of other partners. With additional money from the State of Connecticut, we expanded the program the following year. Farmers were happy, and low-income moms were coming in droves to farmers' markets to buy produce with their special coupons.

By 1989, at least ten states scattered across the country had piloted some version of this program, with each one putting its own unique spin on it. At first we thought the program would be only a slightly offbeat New England innovation when Vermont joined Massachusetts and Connecticut. Like its New England sister states, Vermont had a strong interest in serving low-income families and administered its farmers' market nutrition program out of the state's antipoverty agency. But then Iowa's Department of Agriculture jumped on board as well, which gave the movement credibility with Big Agriculture. Under the direction of Dan Cooper, the only martini-sipping farmer I've ever met from Iowa, the program became a critical part of that state's agricultural diversification strategy. If Iowa's corn and soybean growers were going to survive, the argument went, they had to also start growing fruits and vegetables for their own people instead of importing them all from California. And then New York State joined the program, which, due to the size of its low-income population, a strong agricultural sector, and New York City's Greenmarkets, gave the farmers' market nutrition program the momentum it needed to become a national model. Leaders such as Bob Lewis in the New York State Department of Agriculture and Markets would put so much of their heart and soul into the program that it would eventually become one of the most exciting agriculture and nutrition developments of the late twentieth century.

Later, with the leadership of Gus Schumacher, representatives of the pioneering farmer's market nutrition program states organized themselves into a group that lobbied Congress to start a national pilot program. As a result of some deft lobbying, aided in large part by key contacts that the various participating states had in Congress, a national three-year, $2 million per year pilot program was approved and implemented by the USDA in 1989.

Ironically, this was the first program in the history of the USDA to make a direct connection between farmers, locally produced food, and the specific nutritional needs (fresh fruits and vegetables) of lower-income families. By the late 1990s, Congress was funding the Farmers' Market Nutrition Program (FMNP) at the annual level of $25 million. And in what would become the program's high-water mark, Schumacher, who had moved on to become an assistant secretary of agriculture at the USDA, established the Senior Farmers' Market Nutrition Program (SFMNP) in the waning days of the Clinton administration. This program operated in a similar fashion to the FMNP, but for lower-income senior citizens, and was funded at $15 million per year.

From the work we did in Washington, D.C., a program that had been piloted by activist nonprofit organizations and progressive state agriculture and health departments became a national program funded at $40 million annually. While that amount of money is considered chump change in Washington, it was enough to promote the development and growth of between 1,000 and 2,000 farmers' markets serving low-income households around the country. Between state and federal funding, Connecticut alone has been distributing around $600,000 per year to lower-income women, children, and seniors and has expanded the number of farmers' markets from sixteen in 1987 to more than eighty today. Since the Connecticut program's inception in 1987, it is estimated that it has generated more than $10 million in sales for the state's farmers.

None of this would have happened without public policy and without the corresponding nexus of political and personal relationships that make government go around. Farmers' markets would not have become available in low-income communities without good ideas springing up at the grass roots, then being nurtured by forward-thinking public officials, shared across state lines, and ultimately funded by state and federal government. Without somebody getting up the courage to ask, and without somebody like Ken Anderson being willing to take the first steps, we would be stuck with the status quo.

FROM COTTON TO COLLARDS: ALABAMA'S FARMERS' MARKETS LEAD THE WAY TO BETTER HEALTH

Community supported agriculture, community gardening, and farmers' markets have emerged over the past thirty years as the primary, community-based strategies by which lower-income communities can obtain affordable, locally produced, and, in some cases, organic food. But the relatively small size, availability, and affordability of these projects necessarily limit their impact on low-income households. While farmers cannot be expected to offer their food at discounted prices, government agencies can support low-income participation with subsidies, and nonprofit organizations can develop innovative approaches to the problem. But in light of the size and scope of obesity and food insecurity in this country, can these subsidies and new models be enough to make a difference? One example of a small public policy intervention that has made a difference is the FMNP, which currently does more than any other public or private effort to connect low-income families with affordable, locally grown food. And one of the best stories about how this program has helped both farmers and low-income families can be found far south of the Mason-Dixon Line in the state of Alabama.

Dr. Jill Foster was a physician practicing family medicine in Cincinnati when she became increasingly frustrated treating illnesses that were preventable with a healthy diet. "A young female patient of mine who weighed two hundred pounds asked me, 'Doctor, am I obese?' When I told her she was, the poor child was devastated." Growing tired of situations like these, Foster decided that she needed more nutrition training, something that is sadly lacking in most medical schools today. So rather than endure the quickening torrent of diet-related disease, she chose to study nutritional science in Birmingham, Alabama.

Had she been looking for the fastest route to the belly of the beast, Foster couldn't have done better than Alabama. According to the Trust for America's Health, the state has the highest level of adult obesity (28.4 percent) in the nation. For African Americans, the numbers are worse. Nearly a third of black men and 37 percent of black women are obese. And 440,000 Alabamians, or about 10 percent of the state's population, have been diagnosed with diabetes, a rate that has climbed by more than 50 percent since 1994.

In Birmingham, Foster, a petite black woman, soon noticed that most of the people around her were at least fifty pounds heavier than she was. "Poverty has a lot to do with it," she told me, "and so does race. When you're poor, you eat what's cheap and what's available. In many Birmingham neighborhoods, the only vegetables you'll find are fried okra and fried green tomatoes."

Although Foster understands the roles that high fruit and vegetable prices and limited access play in promoting obesity, she also recognizes the important role that culture plays. One Sunday, for instance, the pastor of her church invited her to share healthy eating tips with the congregation. "As soon as I was done," she said with dismay, "the pastor returned to the pulpit and proceeded to remind everyone about the pig roast dinner on Wednesday and the macaroni and cheese supper on Thursday. The black church is a powerful keeper of traditions, some of which aren't very healthy."

There's another Alabama tradition, however, that may be one antidote to the state's obesity crisis. Drawing on its agrarian roots, Alabama's farmers, community health providers, government agencies, and everyday citizens are starting farmers' markets across the state faster than you can eat a Chilton County peach. While farmers' markets are often viewed in many parts of the country as destinations for willowy food faddists in search of organic yuppie chow, Alabama is looking to them as a way to connect people to sources of healthy, locally grown food. In fact, Alabama may have the fastest farmers' market growth rate in the country. From 17 locations in 1999 to 90 today, the state now has at least one farmers' market in every one of its 67 counties.

It's 93 degrees in Greene County on a June morning in 2006. The occasional breeze isn't enough to stir the feathery blossoms of the mimosas, but the suffocating heat under the metal-roofed farmers' market shed doesn't stop Eugene Hall's frenetic pacing. A contagiously enthusiastic African American farmer, Hall is excited about developing opportunities for Alabama's small farmers and the possibility that they might renew impoverished rural communities is palpable. With his string bean–thin partner, Willie Busby, and a small grant from the State of Alabama, Hall and the Federation of Southern Cooperatives spearheaded the development of this farmers' market. "Build it, and they will come," says Hall, echoing the now famous *Field of Dreams* line. Here in Alabama, the field isn't Iowa corn but hundreds of small watermelon and collard patches, and the dream is to stanch the outmigration of young people and reduce the devastating impact of unhealthy eating.

"The market shed can be one thing that helps revive this community," Hall asserts. By attracting ten or so farmers to this location several days a week, he and Busby are providing expanded market opportunities for local growers and "keeping local money in the local area." In addition to farming, he now works with the USDA Natural Resources Conservation Service, which received $1.5 million to help Alabama's farmers install plastic mulch and drip irrigation systems. "With this simple technology," notes Hall, "farmers can double their yields. They can plant watermelons in the spring, harvest them in late August, and then plant collards in the same place for a late-fall harvest. You're making more money and eliminating most of the backbreaking labor."

But Greene County is poor, and its population is dispersed across a sparse rural landscape. So with $17,000 in farmers' market coupons provided by the Alabama Farmers Market Authority (FMA), lower-income mothers, children, and elderly people have been given an incentive to buy fresh fruits and vegetables at the Greene County farmers' market. "Government money is being used to promote something we all need—food," says Hall, and in this case it is locally grown fruits and vegetables, which are sorely needed by local households.

Don Wambles, executive director of the FMA, doesn't come across at first as the apotheosis of progressive government. A cotton and peanut farmer for twenty years and a former Republican county commissioner, he has a good old boy drawl that strains the ears of his Yankee interviewer. But as the universally acknowledged father of the Alabama farmers' markets movement, he has led a quiet revolution that is signaling a new direction for agriculture not just in Alabama but nationwide.

"Over the course of my farming career, I'd have one good year in four," Wambles says. Like most other commodity crop farmers, he continued to be paid the same prices for his products even as his input costs—fertilizer, equipment, gasoline—went up. Although he had one of the biggest farms in Pike County, the bank told him that he had to get bigger or get out. "I was tired of fighting a losing battle, so I got out of farming and took the FMA job in 1995," he says. Based on the hard lessons he learned as a "price taker" in the cotton and peanut markets, he decided to push his fellow farmers toward something they could have more control over—namely, fruit and vegetable production for direct sale to consumers.

But the push wasn't enough. Wambles also needed a "pull," which he discovered in the form of the federally funded FMNP. The program, which Wambles brought to Alabama, made immediate sense to him when considered in terms of its economic benefits to farmers. For personal reasons, he later realized its value to low-income families as well. "I grew up real poor," he says. "My parents divorced when I was in first grade, and my mother raised two boys on $3,672 per year plus $50 per month in child support from the State of Alabama." He adds that he also studies the Bible and "it tells us to take care of one another."

In 2001, Wambles aggressively pursued a related pot of federal funds, the Senior FMNP, which he credits with accelerating the growth in farmers' markets across the state. In 2005, Wambles and his partner agencies distributed $1.3 million in coupons to about 80,000 lower-income seniors, mothers, and children, who consumed hundreds of thousands of pounds of fresh fruits and vegetables that they wouldn't have otherwise. In turn, more than 1,300 farmers started making serious money at farmers' markets and slowly shifting their acreage from cotton to vegetables.

When it comes to farmers' markets, Art Sessions was one of the state's first big farmers to get religion. Growing pecans, cotton, and some fruits and vegetables, he relied exclusively on wholesale markets to sell his crops. But after giving the new Mobile farmers' market a try, he reduced his wholesale commodity crop activity and increased his share of retail fruit and vegetable production. "Cotton might take a big hit in the next farm bill," Sessions says, referring to the growing reluctance on the part of Congress to provide cotton subsidies. "Over 25 percent of my vegetable sales now come from senior citizens and WIC moms, and many of them come back after they've used up their coupons. I must say that the FMNP is one of the few federal programs that really works."

Danya Walls and Mamie Pollard are two women who don't know each other and, at first glance, seem to have little in common. Walls is a white, twentysomething mother of one child who lives in rural Blount County, about an hour north of Birmingham. Pollard is a seventy-four-year-old African American widow who lives in Birmingham and has eight grown children. In spite of their outward differences, they both participate in the FMNP—Walls as a WIC mom and Pollard as a senior.

Reflecting on Alabama's high diabetes rate, Walls says, "There's a lot of fried southern food and a profound lack of education about healthy cooking." She says that fruits and vegetables are a regular part of her and her son's diet and credits *The Oprah Winfrey Show* with teaching her about healthy food. Pollard has a similar take on the problem. She says, "Children didn't have those kinds of diseases [diabetes] long ago. The way today's parents are feeding their children is wrong—taking them to Mickey D's or Pizza Hut instead of preparing a proper meal."

Walls credits the FMNP with increasing her household's consumption of fresh fruits and vegetables. "We'll eat tomatoes and cantaloupe five out of seven days," she says. For Pollard, the $20 in FMNP coupons she receives is a big help. Her Social Security check is small, and after copayments for medicine, "there's not much left over for fruits and vegetables."

But there's another surprising factor that unites these two women, both of whom must make every penny count: they have a strong desire to help farmers. "Going to the farmers' market is better than going to the grocery store because the food is better and cheaper," Walls says. "I know the farmers are tight on money, so I just want to help them. At the farmers' market, I always try to buy something from everyone."

"Yes, we're blessed to have this opportunity to go to the farmers' market," Pollard explains. "At the store, they spray the food with something, and I trust the farmer more. But the Alabama farmers are having a hard time, and they need our help. That's why we spend our coupons with them."

Surveys of FMNP recipients by state agencies uniformly show that the coupons stimulate increased consumption of fresh fruits and vegetables. In Dr. Jill Foster's opinion, this is one important way to attack the nation's obesity problem, which according to some researchers will make this generation of children the first to have a shorter life span than their parents. "We have to look at the economics of food choices," Foster asserts, "and we have to make it easy and affordable to buy fresh fruits and vegetables."

With only about $1.3 million to spend each year on this program, Don Wambles knows he can't go much further until Congress puts more money into the FMNP. Right now, he's reaching only 20 percent of the state's WIC participants and 27 percent of the low-income seniors. If funding was increased to $7 million, the state's FMNP could give 200,000 of the state's most nutritionally vulnerable people a strong incentive to buy fresh fruits and vegetables.

That figure of $7 million is less than 11 percent of the $66 million in cotton subsidies that went to 8,500 Alabama growers in 2004. Given that the state spends $1.3 *billion* annually on obesity-related medical expenses, Alabama's consumers and taxpayers could potentially get a much higher rate of return from improved health outcomes if there was a greater investment in the FMNP.

Farmers would win, too. As Eugene Hall says, "I can plant one hundred acres of cotton and be guaranteed a price [because of the subsidy], but there's no way to guarantee a price to one hundred farmers who each plant one acre of collards." A robust FMNP wouldn't provide a guaranteed price, but it would stimulate more Alabama farmers to produce food for their own communities, fuel the growth in farmers' markets, and give Alabama's lower-income consumers access to the healthy and affordable food they so desperately need.

CONNECTICUT'S FOOD POLICY COUNCILS

The more people seek to gain control over the food they eat and their larger food system, the more they discover the intrinsic logic of addressing food issues and increasing the food security of their own communities at the local and state levels. Because of their interest in community food security, activists, parents, farmers, and anyone who feels that he or she has a dog in the fight for healthy and affordable food have been turning increasingly to food policy and food system planning.

There is a long and growing list of opportunities for fruitful government cooperation and action on behalf of the food system. But to maximize the potential role of local and state government in addressing food needs, a number of obstacles must be overcome. For instance, not a single U.S. city has a "department of food." A few state departments of agriculture have the word *food* in their names, but rarely does their authority extend beyond the production end of the food chain. Planning departments do not as yet generally recognize the food system as a unique area of concern requiring special attention. As Kami Pothukuchi and Jerry Kaufman pointed out in their 2000 article "The Food System: A Stranger to the Planning Field," "The food system is notable by its absence from the writings of planning scholars, from the plans prepared by planning practitioners, and from the classrooms in which planning students are taught." Because of the work of Pothukuchi and Kaufman, the American Planning Association is considering a proposal to make food system planning an integral part of the planning profession.

The size, scope, and complexity of the food system have often frustrated those who have tried to identify a consistent role for both planning and policy. Yet when the analysis is done, the need for a comprehensive approach to food policy becomes apparent. In a 1996 report to the Connecticut legislature titled "Toward Food Security for Connecticut," the legislature's Planning and Development Committee wrote:

> Sufficient cause exists for Connecticut policy makers to give the security of the state's food system a critical look to ensure that a safe, affordable, and quality food supply is available to all, both now and in the future. The state's food system is large, complex, and paradoxical. The food industry is a substantial contributor to the state's economy, yet hunger, malnutrition, and limited access to food for the poor are well-documented, forcing publicly funded food assistance programs and private emergency food sources to play an ever increasing role in feeding the poor. Connecticut has experienced unprecedented growth since World War Two, but that growth has been accompanied by a corresponding decline in farmland and an increase in environmental degradation. State government provides more services and plays a larger role in everyone's life, but the state agencies that address the production, distribution, and quality of food rarely coordinate their efforts.

This report led directly to the legislature's creation of the Connecticut Food Policy Council. With the organization of the council the following year, Connecticut had for the first time a single entity that could examine the performance of the state's food system and make recommendations, particularly at the government level, for improvements. While it is not a department of food per se, the council is the one place where every aspect of the food system is considered.

In fact, the best examples so far of food system planning and local and state food policy may be in Connecticut. Hartford was among the first cities in the country to have a municipal food policy council, while Connecticut was the first state to have a state government food policy council. In addition, the Capitol Region Council of Governments, the regional planning agency for the thirty-two-town Hartford area, was one of the first such agencies to make food system planning a part of its regional development plan. Both the city and state food policy councils grew out of the work of the Hartford Food System, and the regional planning effort also was inspired and influenced by the Hartford Food System. Let's take a brief look at how these three entities have used their respective vehicles to conduct food system planning and affect food policy.

Hartford Food Policy Council

Founded in 1992, the Hartford Food Policy Council was formed by a city ordinance that established both a city food policy and a council whose task it was to advise city government on how that policy should be implemented. As stated in the ordinance, "The purpose of the policy shall be to integrate all agencies of the city in a common effort to improve the availability of safe and nutritious food at reasonable prices for all residents, particularly those in need." The policy statement goes on to identify fourteen functions of city government that will be used to implement this policy. They include transportation, land use, direct services (food assistance programs such as WIC and school meals, which are administered by city agencies), education, health inspections, and business development.

The council, whose members are appointed by the city council and include representatives of several city departments, monitors the performance of the city's food system, researches and analyzes food issues, and works to improve city government's response to food problems. In chapter six I discussed the council's efforts on behalf of public transportation for low-income residents to travel to supermarkets, and in chapter seven I explained its work in removing soft drink machines from the city's schools. Another example pertaining to the operation of the WIC Program by the city's health department illustrates the critical role that local advocates must play to ensure that federal programs such as WIC deliver the services they are supposed to. As the following story suggests, sometimes food policy work is not very glorious, and in fact it can be downright mundane when you have to slog through government bureaucracy to prevent people from falling into the food gap.

In 2001 Magdalena, a young, low-income mother living in Hartford's Southend, called the city health department's WIC clinic to find out how she could get some additional food for herself and her two-year-old child. The phone at the clinic rang, and rang, and rang. Nobody picked up. She tried again and got the same response. She learned from a friend where the clinic was located and decided to visit it when she got off her part-time job at the nearby McDonald's. Magdalena found her way to the Park Street clinic only to discover that it was closed—not just at that moment but apparently for the past six months due to a fire in the clinic's kitchen. A small, handwritten notice taped to the inside of the window told anyone wanting WIC services to call the same number she'd been trying to call for three days. The notice also said that interested people could visit the city's other clinic in the Northend.

The following day was Magdalena's day off. She made arrangements for her child to stay with a friend and then caught two buses to the Northend clinic, where she arrived just a few minutes before noon. The door was locked, and nobody answered when she pushed the buzzer. She noticed a small sign next to the door advising her that the clinic was closed for lunch between 12:00 and 1:00. The sign also told her that there were no office hours after 4:30 or on the weekend. Since her friend had to be at work by 2:00, Magdalena couldn't wait and had to take the bus home to retrieve her child. The whole trip had been a total waste of time. She was exhausted, frustrated, and just as food insecure as she had been four days before. A program run by the City of Hartford Health Department was failing to do its job and, through negligence, leaving hundreds of low-income mothers out in the cold.

Once the Hartford Food Policy Council became aware of this situation (the council had been informed secretly by a member of the WIC staff about how dysfunctional the program had become), it began to investigate. The council discovered that the WIC office was understaffed because the city's hiring procedures were so convoluted that it required months to fill even the most rudimentary positions. In addition, low staff morale, combined with a host of labor grievances related to highly rigid work rules, had predisposed existing staff to ignore phone calls. The Park Street clinic, which served the entire southern half of the city, remained closed because the city building inspector had deemed the office unfit for use even though the fire was minor and confined to one small part of the building. The city's complex bidding procedures and simple neglect had slowed the repair process to a crawl. The Northend clinic, as Magdalena had discovered, was several miles away and difficult to reach by public transportation. As a result of incompetence, willful neglect, and an intransigent city bureaucracy, the number of WIC participants had plunged in one year from 11,000 to 6,000. Literally thousands of the city's most vulnerable people—mothers, newborn infants, and young children—had been vigorously thwarted in their attempts to get help.

After the food policy council reported its findings to the city manager, mayor, and members of the city council, the city convened a task force to work on the problem immediately. It investigated the labor issues, office problems, and client services practices. Within six months things began to change. Vacant positions were filled, and the phones were answered. The Park Street office was reopened. Clinic hours were extended into the early evening one day a week, and Saturday morning hours were added. These extended hours were necessary because changes in welfare rules had sent many low-income, WIC-eligible women into the work force, which meant that they could not visit the clinic during normal business hours.

Complex but essential food system services such as WIC don't always perform well unless somebody cares about and understands how they work. The Hartford Food Policy Council, acting as the unofficial city food system ombudsman, had the resolve and the expertise to make this federally mandated program function effectively at the local level.

Connecticut Food Policy Council

The Connecticut Food Policy Council was created by the Connecticut legislature in 1997. The state statute charges the council to "develop, coordinate and implement a food system policy linking economic development, environmental protection and preservation with farming and urban issues." Its members include representatives from six state agencies and from six nongovernmental food sectors (the latter group is appointed by the legislature). Like the Hartford council, the Connecticut Food Policy Council monitors, analyzes, and advises, but it also takes an active role in public education. One of the best examples of this was a statewide conference on farmland loss sponsored by the council. This conference led to the creation of a farmland preservation advocacy group, the Working Lands Alliance, and a private land trust, the Connecticut Farmland Trust. Along with the council, both organizations were highly successful in securing enormous increases in support for farmland preservation efforts.

The council's coordinating work has led to the creation of numerous endeavors that address nutrition and health, agriculture, and hunger. One example is the Connecticut Farm Map, which highlights locations where consumers can buy agricultural products directly from farmers. This project came about as the result of two council agency members, the Department of Transportation and the Department of Agriculture, agreeing to work together on a joint effort. By promoting similar cooperative efforts—doing nothing more complicated than putting two people together at the same table who don't normally work together—the council has been able to achieve a number of other significant breakthroughs. The Connecticut

Department of Corrections, the state's single biggest food customer, began purchasing food from Connecticut farmers. The Department of Social Services, which runs the Food Stamp Program in Connecticut, agreed to start bringing EBT technology to farmers' markets to enable farmers to receive food stamps from their customers. Several state agencies, which previously operated independent nutrition education programs, agreed to collaborate with one another by creating a common website for nutrition information and services. Perhaps the Connecticut Food Policy Council's most important role is to serve as a place where the state's food system experts and representatives, in both the private and public sectors, can sit down together, develop mutual trust, and identify where the food system needs strengthening. Some of the best policy work that we did in Connecticut occurred when people stepped outside of their silos.

Capital Region Council of Governments

The most interesting development for food system planning may lie with the Capitol Region Council of Governments (CRCOG), which included two sections on food system issues in its 2003 report "Achieving the Balance: A Plan of Conservation and Development for the Capitol Region." These two sections, titled "Open Space and Farmland Preservation" and "Food System," identified the food and farming challenges facing the Capitol Region, including the preservation of farmland, the development and maintenance of transportation systems that allow all residents access to healthy food, the public health consequences of unhealthy diets, and the composting of food waste.

This seminal work on food system planning drew on previous efforts to highlight these issues, including the work of the city and state food policy councils. It also built on the "Connecticut State Plan of Conservation and Development," which provides general guidance to the state's 169 municipalities on how to manage their growth. The state plan for 1998 to 2003 stressed the need for municipalities to preserve farmland and its long-term food production capability (skills, markets, infrastructure, natural resources) in order to ensure food security. The CRCOG document applied these recommendations to the thirty-two-town Capitol Region. Here again, a public agency stepped outside its routine planning box and in effect said, "How can we integrate food and farming needs into our work? Planning shouldn't just be about lines on a map, it should also be about the basic necessities that contribute to our region's food security."

LOCAL AND STATE FOOD POLICY: WHERE THE ACTION IS

It is in state legislatures and city councils, county boards and planning commissions, and the day-to-day to-and-fro that is community life that decisions affecting our food system are increasingly being made. What has transpired in the past ten years through local and state government with respect to food policy has been enormous; what could take place in the future is unlimited. A good part of the reason that there is so much potential to create policies promoting food security, local food systems, and economic justice at the local and state levels is that this is the arena in which people and small, local organizations participate. Not to negate the importance of the nation's Food Stamp Program, but more people, at least in terms of raw numbers, will support a state bill to ban junk food in schools than will weigh in from across the country to support a piece of national food stamp legislation.

One reason that national anti-hunger policies may stir so little interest is disempowerment of the poor. As New York City's Joel Berg has said, "Politicians spend the last week of their election campaigns appealing to lower-income voters and their entire term in office ignoring them." Without a continual drumbeat of public concern and mobilization by those directly affected by food and poverty programs, few politicians will advocate aggressively to end poverty and hunger. And elected officials also have told me over the years that they don't hear from the poor, a group that traditionally has low voter turnout.

Community apathy toward poverty contributes to low public support for anti-hunger programs. The evolution of our social welfare system from a down-home, albeit stingy and arbitrary, form of charity to one that is highly institutionalized, bureaucratized, and effectively removed from the sight of most Americans has changed our perception of human need. The contemporary Depression-era reader of John Steinbeck's *Grapes of Wrath* could not help but feel compassion for the hungry, the homeless, and the plight of rural America after reading this book. But when social problems such as hunger and poverty, and our confusing national response to them, are as abstract as they are today, it is easy to unplug our hearts and minds from them. And when empathy or compassion is anemic, it is easier to abdicate responsibility for solving those problems to the welfare bureaucracy or some generalized form of charity such as food banks or institutions of faith.

Social science research has been helpful in elucidating the relationships between community participation and problem solving. In the community development field, for instance, where the objective is to address not just a place's housing and economic infrastructure but also its social structure, research has found that the most successful community development organizations are those that connect their residents to one another as well as to their programs. In what has become a sociological classic in a very short period of time, Harvard sociologist Robert Putnam's *Bowling Alone* makes a strong case for the role that social capital (also known as social networks) plays in a community's health and well-being.

Taking a cue from this social science research, the Hartford Food System in 2001 supported a study by Katie S. Martin that examined the relationship between food security/insecurity and social networks. Based on face-to-face interviews with 330 lower-income Hartford families, we found the following:

- Low-income families were more likely to be food secure if their social capital—that is, their connection to local social networks—was high.
- A high percentage of food insecure families do not participate in food programs: 45 percent did not receive food stamps, 67 percent did not use food pantries, and 37 percent who were eligible for the WIC Program did not participate in it.
- Lack of access to large supermarkets and/or transportation to get to large supermarkets was significantly associated with food insecurity. Fifty-five percent of the respondents did not have a car.

What emerged loud and clear from this two-year study was that social networks and participation in community life, or the lack thereof, could either reduce or exacerbate the ill effects of poverty.

This is why people across the country are focusing on local and state action to address some of their most entrenched social problems. According to Rodger Cooley, the domestic policy director for Heifer International, an Arkansas-based nonprofit that works on global and U.S. food issues, there has been a definite shift away from the word *hunger*, with its implication that we simply need to distribute more food to people, to the terms *food security* and *community food security* as the guiding principles for Heifer's work domestically and internationally. While emergency food distributions continue at high levels to "feed the hungry," more food system stakeholders are now saying that the focus should be on improving access to good-quality, affordable food, especially among low-income families, than only to increase the overall quantity of food available. Efforts to bring supermarkets to lower-income urban neighborhoods, such as those in Chicago spearheaded by community food activist LaDonna Redmond, are one indication that the people who live in these neighborhoods want a normal food system, not one dependent on charity.

The shift to a community food security framework also has contributed to the growth in community gardens, farmers' markets, and CSAs that assist low-income communities. Overall, an informal alliance between sustainable agriculture and food security advocates has emerged that shows promise for helping both the poor and small and medium-size farmers.

The City of Chicago has placed food security on its policy agenda and is putting its massive food-purchasing power behind efforts to strengthen Illinois's small farm sector. In cooperation with the Chicago Food System Council, led by Erika Allen and Rodger Cooley, Chicago city government is using its resources to expand urban agriculture (one feature of which is improving municipal composting), food retailing, and the purchasing of regionally produced food by the city's schools and other public institutions. And in what may be a first for any city, Chicago issued a position paper on the 2007 U.S. farm bill articulating its interest in the future of national food and farm policy.

FROM WINGTIPS AND NECKTIES TO BOOTS AND BOLOS

My first visit to the New Mexico legislative building shortly after I moved to Santa Fe four years ago was nothing short of traumatic. Having been used to the tweedy, buttoned-down look of most Connecticut legislators, I thought I had mistakenly walked into a cattle roundup. There were no wingtips or Brooks Brothers suits here. Instead, the men were attired in cowboy boots, blue jeans, bolo ties, and large black cowboy hats. Though dressed in slightly more formal garb, the women wore leather-fringed vests, jackets, and skirts; more feminine versions of cowboy hats and boots; and enough silver and turquoise jewelry to send a metal detector into red alert. Several women wore belt buckles so thick that they could stop a .45-caliber slug at twenty paces. Their flamboyant dress and accessories reminded me of the Mae West admonition, "It's better to be looked over than overlooked."

With the same trepidation that I experienced during my early policy years in Connecticut, I wondered how I could ever expect to influence food policy in a place like this. Not only did a person have to know how to round up cattle with a horse, but it wouldn't hurt to be a rodeo queen as well. Which is why it came as no surprise that Pam Roy, codirector of the Santa Fe–based nonprofit Farm to Table, has been so successful securing food policy victories in the New Mexico legislature. Not only can this petite woman round up cattle astride a 1,200-pound galloping horse, but she also can play polo-cross—a horse-mounted team sport that requires players to hurl and catch a rubber ball with a short-handled lacrosse racket. And, yes, it also turns out that she was the 1977 Santa Fe rodeo queen.

The New Mexico legislature is no less sophisticated than Connecticut's lawmaking body, so horseback riding skills per se will not in fact help a person get laws passed. But the bond that those horse tales provide, to say nothing of the personality required to win the congeniality phase of the rodeo queen contest, gives Roy the inside track on cultivating the relationships she needs to lasso votes. To follow her down the legislative corridors is to see a side of policymaking that never gets covered in civics textbooks.

Having grown up in Santa Fe and run the world-class Santa Fe Farmers Market for ten years, Roy knows just about everyone in New Mexico's capital city. She hugs and kisses her way down the hallways, greeting lobbyists, legislators, and administration officials alike. If she doesn't hug and kiss them, it just means that she hasn't met them yet—in which case she gives them a firm, two-handed handshake that has earned her compliments from six-foot five-inch cowboys.

Roy uses familiarity and respect to open doors, prevailing on legislative leaders and administration officials to pass laws, create task forces, and otherwise focus the state's attention on food, nutrition, and agriculture issues. She could easily have put her skills to work for some of the state's more infamous, higher-paying interests, such as the factory dairy farms or the uranium mines. But Roy and the organization she represents, the New Mexico Food and Agriculture Policy Council, are committed to promoting sustainable farming and closing the state's yawning food gap. And in a state that was ranked number one in food insecurity in 2005; where obesity and diabetes are enormous problems, especially among the state's twenty-two sovereign Native American tribes; and where a fifty-mile drive to a supermarket is not unusual, the food gap is a bear of a problem.

Roy has used more than her lobbying skills to take on the state's deep social and economic problems. She also has used her relationships with dozens of private individuals, organizations, and state agencies to organize the Food and Agriculture Policy Council and, to a lesser extent, the New Mexico Task Force to End Hunger. Over the past five years, these two organizations, which represent virtually every facet of the state's food system, have proposed and worked for a series of policy actions by state government that are reducing hunger, improving access to healthy food, and increasing the viability of New Mexico's small farming sector. Their specific successes include winning, against the vigorous opposition of the soft drink industry, nutrition standards for the state's schools. They have secured state funding for the expansion of the federal School Breakfast Program and an increase in the minimum food stamp benefit to $30 per month (from the current nationwide minimum of $10, which has not changed in thirty years). They also have secured increased funding for small farmer initiatives and developed a program to purchase New Mexico–grown fruits and vegetables for the state's schools and food banks. In addition to their funding achievements, Roy and the Food and Agriculture Policy Council have prevailed upon the legislature to form the Rural Food Gap Task Force. This group comprises the state's economic development and human service agencies, as well as specific food industry representatives, and has been charged with developing initiatives that will bring retail food outlets to underserved New Mexico communities.

Persistence, knowledge of the food system and the legislative process, and a wide network of relationships have been the keys to success for a mounting number of food policy advances in New Mexico. And it doesn't hurt if you also know how to ride a horse.

PROJECTS, PARTNERS, AND POLICY

Three things are necessary to change our food system and close the food gap: projects, partners, and policy. Without these "three Ps," synchronized and fully engaged, we will never be able to develop the innovation, know-how, or resources necessary to reach those goals. *Projects* are singular activities that social justice and local food system advocates pursue—such as farmers' markets, food banks, and improving the delivery of food assistance programs. They are the oil that lubricates the gears, but more important, they are the source of innovation that inspires others to replicate and disseminate them and, as people learn more from their own experience, to innovate again. *Partners* are the nexus of relationships and the wellspring of social capital that we draw from to accomplish our work in today's complex world. No one person, organization, or approach will close the food gap. Doing so will require an extended and long-term team effort with some uncommon connections and often un-thought-of players. But it is *policy* that will "make the right . . . prevalent." Getting our heads above our own projects, or even above our own supper plates, is necessary to see the opportunity that public policy affords. We cannot expect change to occur unless we can replicate everywhere the good work that is going on in specific parts of the country, and to do that requires the broad shoulders of government to push with us in the same direction at the same time.

Policy work is usually understood best and first at the local and state levels, where lawmakers and administrative systems are more accessible. When public resources and infrastructure are harnessed to the same wagon that private activists are pulling, innovation is fostered and diffused more rapidly and widely than it would be otherwise. Although we can learn the policy ropes best at the lower levels of government and use those places to spark modest fires for change, the most profound change will occur at the national level. For that reason, we must continue to learn the ropes in Washington, D.C., as well and not take no for an answer from the public and private gatekeepers who guard the corridors of power so zealously.

CHAPTER TEN

Income Disparities, Poverty, and the Food Gap

An imbalance between rich and poor is the oldest and most fatal ailment of all republics.
Plutarch

A PERSONAL QUESTION that I'm often asked is why I chose food for a career. Why not law, investment banking, or teaching? Those are the professional cloaks that people want to drape me in. To get past the question as quickly as possible, I respond that with food, I do far less harm than I would with those other careers, and after all, I love to eat. But obviously there's more to it than that. One doesn't decide to passionately pursue something that he or she knows as little about as I did about food and farming without having some justification.

For me the answer came early. Because of the idealism of the overheated sixties and the fever inside my head that egged me on to change the world, I came quickly to the conclusion that food was the most tangible and direct way to help individuals and communities gain a measure of control over their lives. Counseling and other healing professions didn't seem to suit my temperament. Whenever I dabbled in those worthy enterprises, I soon found myself focusing my energy and that of my clients on the wrongs that society may have done to them. While that was justified in some cases, it was often an unfortunate projection that prematurely turned the clients away from some of their more fundamental problems. Similarly, I considered politics—not that I could ever imagine in retrospect that anyone would vote for me, but at least it seemed at the time like the most direct way, short of armed revolution, to secure the new world order that I so fervently desired. But alas, like counseling, politics required a kind of patience, discretion, and personality that the good Lord had not blessed me with. Safer, I decided, to clench the handle of a hoe or a sturdy stalk of broccoli than the neck of some knuckle-dragging mayor.

Food was the path to empowerment, if not spiritual enlightenment as well. At a personal level, so-called food work could relieve me of a certain amount of existential torment: if I plant this garden because I have to eat, all other choices will quickly fade to trivial. Reducing life to its simplest components can be a relief for many, especially the young, and especially in a world where an overabundance of choices is no longer an emblem of our freedom but more a sign of our slavery. At the community, regional, national, and global levels, food confronts us, along with air and water, with life's most elemental needs. Our food security, though taken for granted by most Americans today, is still the bedrock of our existence. The earth, the skill to steward it, and the means to process, store, and distribute food can make or break a civilization. A Dutch friend of mine reminded me of what a fragile thing food security is when he told me that during World War II, his father was forced to trade a gold watch for half a loaf of bread.

It was for these reasons and others that I concluded that whatever modest footprint I might leave in this world, it would be in pursuit of efforts to enhance the food security of individuals, communities, and regions. Eliminating poverty, ensuring health care for all, and providing the highest-quality education to every child are all critically important and must be addressed more aggressively in order to achieve food security. Indeed, as my colleague Joel Berg has said, "Poverty and inequality of wealth are the problems of the twenty-first century." But in a country reluctant to attack the root causes of poverty and redistribute the wealth that is ultimately created by each of us, attacking hunger and food insecurity are the best routes available for now. If nothing else, they will connect Americans to short-term solutions and, over time, to the long-term one—namely, fighting poverty.

We have in America today a tale of two food systems—one for the poor and one for everyone else. As discussed throughout this book, lower-income Americans have moved along a spectrum of hunger and food insecurity throughout the history of the country but did not become visible, or at least publicly acknowledged, until the Great Depression. It was at that point, through public leadership and first-of-their-kind government interventions, that organized attempts were made to relieve hunger. But from that point right up to the present, the nation's antihunger policies have always been joined at the hip with attempts to help farmers, promote national security, or serve another interest or constituency. And when, whether by accident or design, the lifeboat got too crowded, it was the hungry who were thrown overboard.

Although the sight or the sense of a person without enough to eat has plucked our moral chords forever, we as a society have never mustered a sustained measure of political gumption to ensure a certain and final end to food insecurity. Just as we have joined the fate of farmers and their crop prices to that of feeding lower-income families, we have decided to serve the needs of malnourished children only if those needs can be tied to national security. In a similar vein, the food industry will make its endless abundance available to food banks only if it can link its waste removal and food surplus utilization needs to its donations. And rather than take the bold steps necessary to end poverty, food insecurity's ultimate cause, we use our large and complex network of public and private antihunger and nutrition programs to manage poverty instead. There is something about the history of addressing hunger and food insecurity in this country that provokes cynicism and frustration. There is always another motive, another deal to be cut, before humanity's needs are met.

Even when it comes to foreign food aid, it is American policy to relieve the starvation of Africans and others only if the donated food is grown by American farmers and shipped on American ships. As a result of the rising costs of U.S. grain and shipping that grain from U.S. ports to Africa, U.S. food funds are buying less food than they did in the past. European countries, Australia, and Canada are moving away from this practice by simply giving money directly to African nations so that they can buy food from more local or regional sources, a practice that also stimulates the development of local agriculture, which in turn promotes regional food security.

Although our national ambivalence toward hunger and poverty slows our progress to a crawl, we show no hesitation about accelerating our pursuit of the dietary good life. The growth in the markets for organic and local produce is nothing short of astounding. One March 2007 *Time* magazine cover story titled "Forget Organic. Eat Local" contained much hand-wringing over the questions of how local is local and how important is organic. (Cover stories like this are one way that our culture bestows legitimacy on a movement.) Of course, the piece made no mention of how affordable either local or organic is, assuming that if you have to ask how much your food costs, you can't afford it.

In an excellent January 2007 article for the *New York Times Magazine* titled "Unhappy Meals," Michael Pollan ably dissected our dietary dilemmas and the games that our mischievous food system plays. He closed the piece with a simple list of action steps that consumers can take to avoid many of the traps of America's food marketing machine. His recommendations included eating real food, eating local food, staying away from supermarkets, and avoiding food whose ingredients we can't pronounce. Good advice. But he also told us to pay more for food, because "there's no escaping the fact that better food—measured by taste or nutritional quality—which often correspond—costs more, because it has been grown or raised less intensively and with more care." He went on to admit, "Not everyone can afford to eat well in America, which is shameful," but he offered no suggestions as to how that shame could be erased.

Paying more for the best doesn't seem to be a particularly tough challenge these days. As the *New York Times* said of Whole Foods Market, it "has built its empire . . . by capitalizing on the willingness of consumers to pay more for organic and natural foods." That Whole Foods blows away the competition by charging more than any other store is not surprising to anyone. One Chicago area retail price survey of twenty-one supermarkets found that Whole Foods, the only "natural" food store on the list, not only was more expensive than any of the other stores but was actually 30 percent higher than the next-highest-priced store.

But rather than worry about the effects of high prices on lower- and even middle-income consumers, Whole Foods has turned its attention to local farmers, in whom, at the behest of Pollan and others, the chain is going to invest $10 million to make them "Whole Foods–ready." While this may strengthen local agriculture and bring the average Whole Foods shopper a wee bit closer to "local food," it will do nothing for the low-income mom who is riding the bus to the Wal-Mart on the outskirts of town. Given that there is a growing shortage of farmers, Whole Foods' actions may even be harmful to low-income interests by causing what I call the "Greenwich effect." As soon as the housewives in very upscale Greenwich, Connecticut, organized a farmers' market, farmers left the hard-pressed urban markets faster than spinach bolting in July. In the same way, the continual push by affluent shoppers and the nation's retail bastions of naturalness to procure local and organic food will only increase prices and widen the food gap between them and lower-income shoppers.

None of this, unfortunately, will help encourage consumers to eat more healthy food. According to the USDA's chief economist, Keith Collins, fresh fruit and vegetable prices rose over the years 2001 to 2006 faster than those for any other food category and are predicted to continue their climb at 4 percent per year. The impact of this, he predicts, will be a reduction in per capita consumption. The impact will also be higher prices for fruits and vegetables—the one item we need to be eating more of—and higher prices for organic and locally produced food as farmers flock to the high-end natural food price gouger. And that will mean fewer sources and higher prices for lower-income consumers who manage to find their way to farmers' markets. We have continued our historic pattern of putting the farmer first while ignoring or segregating the lower-income shopper.

MANAGING POVERTY

There are 26.3 million Americans enrolled in the Food Stamp Program. According to the USDA, total food stamp spending for 2006 was $32.8 billion, the highest level ever for the program. These numbers work out to an average per person allotment of $94.04 per month. If we assume 3 meals per day times 30 days per month, food stamps give the average American half a penny less than $1.05 per meal, or $3.14 per day. Eat hardy and be sure to clean your plate, as my mother would say.

In addition to the woeful inadequacy of the Food Stamp Program, whose benefits are typically exhausted by the third week of the month (which is when food banks get their greatest deluge of customers), a tremendous number of eligible or needy Americans are not well served by the program or served at all. Wading through USDA Economic Research Service analyses of food insecurity, one finds that out of 35 million food insecure Americans in 2005:

- Nearly half of food stamp households and about 40 percent of households that receive free or reduced-cost school lunches or WIC were food insecure.
- About 68 percent of households that obtained emergency food from community food pantries were food insecure.
- Households that received food assistance spent less for food than nonrecipient households—about 90 percent of the Thrifty Food Plan, which is the government's minimal measure of a nutritionally adequate diet, regarded by many as a wholly inadequate food plan.
- Only 55.6 percent of food insecure households received assistance from at least one of the three largest food assistance programs. Of these food insecure households, 35.6 percent received assistance from the Food Stamp Program, 32.8 from the National School Lunch Program, and 12.6 percent from WIC.

To address the low participation rate in the Food Stamp Program—a number that generally hovers between 50 and 60 percent of those who are eligible—the government spends about $5 million annually (which is, coincidentally, almost the same amount that it spends to promote healthy eating). As one who has participated in these brave-hearted attempts to increase participation in the Food Stamp Program, I would have to say, with all due respect to the profound commitment of my colleagues who do this work, that the results are as marginal as the stingy funding that the government provides. Except for some promising efforts to enroll people directly online at food pantries, there is no reason to be satisfied with outreach efforts to date. What can we expect to get for $5 million?

What will the government do about this sorry state of affairs? We can expect no substantive change soon if the U.S. Senate's proposed $2.9 trillion fiscal year 2008 federal budget is any indication. Out of all of this money, there is only $18 billion for nondefense, discretionary spending, an amount that is slightly more than one-half of 1 percent of the total budget. So-called discretionary spending is the only place where new or additional spending initiatives—such as health care and childcare, or nutrition spending—can come from. USDA secretary Mike Johanns has proposed a so-called reform in the eligibility guidelines that would add all of $1.3 billion per year to the Food Stamp Program. Reforms like these, which have been part of the ongoing strategy of antihunger advocates as well as previous administrations, do nothing more than tinker at the margins of America's most important antihunger program. Such actions and budget proposals only reinforce the contention that we are merely managing poverty.

Some advocates have suggested that the government could virtually eliminate food insecurity in the United States by adding $10 per week per food stamp recipient to the program's benefits. This would increase the total cost of the program by almost 50 percent, or about $14 billion. Maybe by the time you read these words, the Iraq War will have ended and the current administration's tax cuts for the rich will have expired. Should these things come to pass and nothing as dire as, say, the bailout of the U.S. auto industry is looming, we may be flush with both cash and a renewed desire to wipe out hunger. But don't hold your breath, because when the line forms for help, you can be sure that the poor will be discreetly ushered to the end.

And in the meantime, don't expect private charity to bail out the poor either. According to the U.S. Conference of Mayors' 2007 annual hunger report, requests for emergency food assistance rose an average of 7 percent in 2006. In addition, "all of the 23 cities in the survey reported that families and individuals are relying on emergency food sources not just in emergencies but as a regular source of food over long periods of time." The mayors' survey found that 23 percent of the demand for emergency food went unmet, that many of the emergency food providers were forced to turn people away, and that the number of families with children asking for emergency food increased in 70 percent of the cities compared to the year before.

THE RICH GET RICHER

Income and educational inequality are clearly at the heart of the food gap. If there is a gap that is more responsible for the food gap than any other, it is this country's glaring disparity between the haves and the have-nots. Economic inequality has been increasing in the United States for more than thirty years. According to the *New York Times*, "The top 0.1 percent of earners—that's one out of every 1,000 families—made 6.8 percent of the nation's pretax income in 2004, up from 4.7 percent a decade earlier and about 2 percent in the '60s and '70s."

As the superrich take a greater share of the national wealth, the poor sink deeper into poverty. The percentage of Americans living in poverty is 11 percent, about 37 million people, and the number of those who live in severe poverty (with incomes that are less than half of the federal poverty level) has reached a thirty-two-year peak. An analysis of the 2005 census by the McClatchy Company found nearly 16 million Americans living in deep or severe poverty, defined as a family of four with an income of $9,903 per year. The number of severely poor persons increased by 26 percent between 2000 and 2005. And just in case people think that the severely poor are a bunch of good-for-nothing working-age men, one in three of the severely poor are children and two in three are women. The McClatchy report assigns at least partial responsibility for the severity of poverty in this country to the U.S. government, which spends a smaller portion of its gross domestic product on antipoverty programs than any other industrialized nation except Mexico and Russia.

Contrast this degree of hardship with the lavish lifestyles of the rich and hoping-to-be-famous. The *New York Times* reported recently on one bat mitzvah that cost $10 million. That much money would actually lift 40,000 severely poor people above the federal poverty level. The investment house of Goldman Sachs gave Christmas bonuses that started at $600,000 to every employee in 2006. Since 47 million Americans do not have any health insurance, including 9 million uninsured children, isn't it possible that more modest bonuses would have allowed Goldman Sachs to provide enough money for thousands of children to receive health coverage? After all, aren't we generous to the rich even when they fail, as was the case for the fired CEO of Home Depot, Robert Nardelli, who received $210 million as a going-away present.

As has been stated elsewhere in this book, Americans keep a tight grasp on the notion that the poor are largely responsible for their plight. The recently deceased sociologist Seymour Martin Lipset made this point more than ten years ago in his book *American Exceptionalism*, noting that 78 percent of Americans endorse the view that "the strength of this country today is mostly based on the success of American business." As other research has borne out, less than a third of all Americans believe that government should intervene to reduce income disparities, and 70 percent believe that "individuals should take more responsibility for providing for themselves." In most European nations and Japan, the numbers are nearly reversed.

Of course, Americans' misperceptions about poverty fly in the face of reality and, even worse, are simply not in our own best interest. We rarely recognize, as Barbara Ehrenreich has done so clearly in her book *Nickel and Dimed*, how dependent we are on low-wage workers, often immigrants, for nearly every facet of our lives. Those who labor for minimum wage or slightly above are the ones who pick our crops, clean our hotel rooms, and bus our tables. Try buying food, going on vacation, or eating out without them. Their poverty and deprivation subsidizes our lifestyles and our businesses. Every low-wage worker living six people to a single room; every immigrant who works at a dangerous, arthritis-inducing slaughterhouse killing our cattle and cutting our meat; every illegal hotel housekeeper who is afraid to go to a food pantry for help or grocery shopping for fear of being deported; and every one of their uninsured and untreated sick children help make America's good life available to the rest of us.

Reinforcing its reputation as a socially stingy country, America also shoots itself in the foot by spending less than any other developed nation on childcare. By making it difficult for ordinary people to obtain childcare, it's difficult for the poor and even the middle class to pursue jobs and careers. And as has been pointed out in some circles, the lack of childcare is contributing to our nation's declining fertility rate. Men and women are putting off having children in order to maintain the two incomes they need to survive.

Plain and simple, poverty is expensive. We as a society can pay now or pay later. In a study by Georgetown University economist Harry J. Holzer, released in January 2007, it was found that children who grow up poor in the United States cost the country $500 billion per year. This is due to the fact that they are less productive, earn less money, commit more crimes, and have more health-related expenses. "The high cost of childhood poverty [which affects 17 percent of all U.S. children] suggests that investing significant resources in poverty reduction might be more cost effective than we thought," said Holzer in testimony before Congress.

As is obvious to those of us who are middle-income or higher, the poor do not live among us. They live where housing is cheap—at the edge of town, in the ghetto, in rural hamlets, trailer parks, and *colonias* along the Mexican border. Out of sight, out of mind. As research on the location of food stores has made clear, these communities, which are characterized by high concentrations of poverty, are not well served by supermarkets and other sources of affordable, nutritious food. If the poor were scattered more or less evenly among higher-income groups, we could surmise, with a fair degree of accuracy, that they would have the same access to healthy and affordable food as the rest of us. They might not be able to go to Whole Foods, but they wouldn't have to struggle to get a good food store. Of course, this is not the case. The poor, with their diminished buying power, are forced to congregate in their own communities, which are spurned by the retail food industry.

CONCLUSION

Resetting America's Table

Do not wait to strike till the iron is hot;
but make it hot by striking.
William Butler Yeats

A DEAR AND RESPECTED MENTOR, Rod MacCrae, the first director of the Toronto Food Policy Council, once told me that he knows much less now that he is in his late fifties than he did when he was in his twenties, which is when, he maintains, he knew everything. I learned many things about organizing, food, and policy from Rod, but humility was not one of them. My community food experiences of the past thirty-five years have been too many, too rich, and too enjoyable to avoid reflection, analysis, and conclusions. But like any garden, conclusions can be fragile or seasonal in their duration. They may be too idiosyncratic, too narrow, or already eclipsed by a younger and more nimble generation that, like Rod and I once did, knows everything. With these caveats so stated, may God grant me the arrogance I need to assert my conclusions, and may He grant me an equal measure of wisdom and humility to accept that others who follow will soon draw their own.

IT'S TIME TO MOVE BEYOND HUNGER

Our society's stated mantra to end hunger has grown tired and hollow. We know its cause—poverty; we know its solution—end poverty. Yet we choose instead to treat hunger only as a symptom of poverty. We are disingenuous by continuing to tinker with the nation's complex, usually disconnected network of food assistance programs, which is zealously defended by a select group of program interests and their associated networks. Similarly, the expansion of food banks and their ongoing solicitation for food will never get us any closer to a solution. They are victims of a concept known as paradoxical counterproductivity, which holds that certain attempts to solve a problem only provide an incentive to make the problem worse. The best metaphor is transportation, where planners attempt to solve increasing traffic congestion with more highway expansion. Within no time at all, more drivers are attracted to those new lanes, which soon become as clogged as the old ones. In the same way, we build bigger and better food banks in response to immediate needs, which only generates more demand and requires yet another round of food bank expansion.

To get us heading down the right road, I think the time is long past due to create a single national food assistance program that works in tandem with an intentional and effective campaign to end poverty. The antipoverty campaign must support health insurance, quality education, childcare, and a living wage for all citizens. The national food assistance program—which I will dub the "Food for All Program"—must be adequately funded to ensure that not only does everyone have enough to eat but also that everyone can afford the same healthy food that many Americans are becoming accustomed to. As the antipoverty campaign begins to reduce poverty, the Food for All Program should be reduced accordingly.

In the same vein, we must seriously examine the role of food banking, which requires that we no longer praise its growth as a sign of our generosity and charity, but instead recognize it as a symbol of our society's failure to hold government accountable for hunger, food insecurity, and poverty. We must begin to reduce the size and scope of food banking in America, as we would with the Food for All Program, in proportion to our success in reducing poverty. Ultimately, food banks should return to their original role of addressing genuine short-term community emergencies, such as natural and man-made disasters. They should cease their role as a dumping ground for the waste and surplus of America's food industry and instead secure and distribute only food that promotes good health. When donations of such food are not available, public funding should be provided to purchase healthy food, with an emphasis on food that is produced locally or regionally. In recognition of the fact that food banking has evolved a vast and sophisticated infrastructure—warehousing, cooling/freezing, trucking—some form of adaptive reuse should be found for food banks that contributes to the long-term food security of their regions, perhaps in the form of helping small and medium-size farmers to develop their marketing capabilities. In the same way, food banks should work as equal partners in close coordination with other community-based and public interest food organizations and agencies to formulate and implement long-term food security plans for their regions.

Just as no quality business enterprise would ever think of operating without a research and development department, neither should important social policy organizations perform their work without the capacity to develop and test new ideas and methods for solving social problems. To that end, we must foster a spirit of innovation, research, and development at the national food and farm policymaking level that actively integrates the ideas, experience, and learning from the local and state levels. The "my way or the highway" approach to antihunger and healthy food advocacy must end. While business as usual may always be the prevailing reality of Washington-based policy making, it doesn't mean that those who are closest to power must accept that reality. If the vision is to end hunger and poverty, a new mindset is required. That mindset must be informed and supported by good practice that comes out of the communities and good theory that comes out of the academies. And those who carry the message to policymakers must gird their loins for battle rather than set the table for accommodation.

WHAT'S GOOD FOR THE AFFLUENT IS GOOD FOR ALL

The educated and well-heeled are flocking to local and organic food in ways that could never have been imagined by even the most hallucinogenic hippie. The phenomenal growth of Whole Foods Market, the explosion of farmers' markets, and the near religious fervor that now guides the local food movement were forces that I would never have anticipated, even when I was standing knee-deep in horse manure in Natick, Massachusetts, thirty-three years ago. These forces speak to the health consciousness that has seized our culture. While some of the science justifying our concerns is still being debated, there is no doubt that our system of industrial food production and marketing has messed with the environment in the same way that it has messed with our health. There can be no empirical equivocation over the disaster that obesity and diabetes have wrought—a disaster that has cut a deeper and wider furrow through the lives of the poor than it has through those of the rich. As diet-related diseases have eclipsed hunger as our nation's biggest food problem and are making inroads as our biggest public health problem, the headlong pursuit of organic and local food has opened up the food gap to a record width.

Unless we are prepared to tolerate two very different food systems, one that serves an elite class very well and one that serves all others poorly, I recommend that we invest our public and private charitable dollars in healthy food at every opportunity. The cost of healthy food, which should include local and organic whenever practicable, should not be a limitation for any class of citizens. Our National School Lunch Program serves more than 29 million children a meal every day. Half of those children are receiving a free meal courtesy of the federal government. Just because it is "free" food doesn't mean it should be cheap food. It should be the best that the richest nation on earth can provide. Cutting-edge "lunch ladies" such as Ann Cooper, the pioneering food service director of the Berkeley, California, schools, and outside-the-box thinkers such as Alice Waters have demonstrated that local, sustainable, and delicious food can be a part of every school cafeteria.

But it also has to be paid for. As my Santa Fe colleague Lynn Walters has said, "School food is a measure of how we value our kids." Schools should not have to cut corners to serve healthful meals. Schools should pay their food service staff a living wage and remind them that, given the magnitude of the threat of poor diets to our nation's children, they may have the most important job in the school. They also should purchase local and/or sustainably produced food whenever possible.

As we do in the schools, we should ensure in every community that geography, income, and race are not barriers to securing a healthy and affordable diet. To that end, we must develop and implement a national program that I will dub "Re-Storing America's Food Deserts." The fact that those with money and cars shop at Whole Foods and the poor catch a bus or two to the local Dread-Mart is unacceptable. There are so many good models of community economic development that have fostered dramatic improvements in local food retail landscapes that it is unnecessary to search very far for answers. What is needed is money. An investment in bringing high-quality retail food outlets, which can include public markets and farmers' markets, to the country's underserved rural and urban communities should be our highest economic development priority. Not only will it pay off in new property taxes and jobs for those communities, but it will lower the nation's health bill as well. And if the retail food industry is expecting only a nice juicy carrot to bring it back to the places that, like an irresponsible father, it has abandoned over the course of the past thirty years, they should think again. There should be a stick as well, one modeled on the 1977 Community Reinvestment Act, which effectively ended the banking industry's practice of redlining poor and minority neighborhoods. If the private sector needs financial help from the public sector to do what it should have been doing all along, it must behave in an ethical and responsible way toward those places that need it the most.

WHAT'S GOOD FOR FARMERS MUST ALSO BE GOOD FOR CONSUMERS

I'm a good gardener and have been a mediocre farmer, both of which have imbued me with a fair degree of empathy for those who till the soil for a living. For reasons that may seem odd to some, I love and adore most of the farmers I have met in my lifetime, even a few who operate some of the nastiest industrial enterprises in our food system. But our national farm policies are now, as they largely have been for the better part of a century, woefully out of balance. They favor large commodity farmers whose contributions to our food supply have damaged both our environment and our health. But in addition to policies that support unsustainable and deleterious practices, there is among many farm and food advocates an attitude of pastoral sentimentality that bestows a status of entitlement on farmers. When taken to a more extreme end, which is not uncommon, this sentimentality fosters a permissive environment that tells the farmer that he or she may charge whatever the market can bear.

I will not defend to the death the right of farmers to charge whatever they think they deserve, anymore than I will vigorously defend policies that have allowed unlimited and questionable commodity crop subsidies for what seems like an eternity. We must have food and farm policies in this country that are just as good for consumers as they are for farmers. If farmers are indeed stewards of the earth, hold a trust that we as a nation have granted them, and are entitled to earn a living wage for their toil, we must pay food prices that serve those needs in return for a commitment from them to produce healthy food for all. As another of my Toronto colleagues, Wayne Roberts, has said, "The right of producers to a good income doesn't override the right of poor consumers to eat. Nor can the need for cheap food override the rights of producers to make a living, or pressure producers to violate the carrying capacity of nature." Policies that balance these needs must be found. Serving one interest at the expense of another is, like the way we've treated the environment, a recipe for disaster.

Our national food and farm policies must be reoriented toward securing a healthy and affordable diet for all. This is not a matter of semantics but a matter of who comes first, the farmer or the consumer. In light of what we know about the relationship between health and food, and between health and the environment, both of which we know more about now than we did thirty years ago, our system of food production should put the consumer first. This in no way implies a cheap food supply. To the contrary; it explicitly means that we, the consumers, must accept the cost that healthy food and a healthy environment imply.

This means that we must be prepared to provide subsidies to those who cannot afford the true cost of healthy, sustainably produced food. One of the best small programs in the federal government is the Farmers' Market Nutrition Program (FMNP), discussed at several points in this book. It is currently funded at $35 million. With $1 billion dollars, an amount that is 5 percent of the federal crop subsidies paid out in 2005 and 2 percent of the total budget for federal food assistance programs in 2006, we could give a substantial and meaningful incentive to all low-income women, children, and senior citizens to buy fresh local produce at farmers' markets. In addition to helping younger people in particular develop a lifetime of good eating habits, it would provide a significant market incentive for fruit and vegetable producers.

At the same time that our farm and food policies reorient the relationship between farmer and consumer, we must invest in a new generation of farmers. Who will grow our food in the future is not an idle question. It is a real problem for which there is currently no good answer. Young farmers, new entry farmers, immigrant farmers, minority farmers, and women farmers are all potential food producers who have special needs for assistance but who may very well be the ones who feed us for the balance of this century. If we are going to subsidize their entry into farming, we should not be doing it only to feed the elite customers of Whole Foods Market.

That's like publicly supporting medical school for doctors whose future practice will be limited to cosmetic surgery for the Greenwich, Connecticut, tennis set instead of basic health care in Harlem.

CREATING FOOD-COMPETENT CITIZENS AND COMMUNITIES

As both a food activist and a parent, I have observed the enormous difference that contact with real food, the soil, and a vegetable plant can make in the lives of young people. Perhaps the best evidence I can present for my position is my own two children, now in their twenties, who, unbeknownst to them and in violation of all human subject protocols, were forced to participate in a lifelong experiment directed by their diabolical father. They gardened, cooked, visited farmers' markets, "volunteered" at our CSA, and over dinner were subjected to my tirades against the evils of the industrial food system. They complained bitterly that unlike their friends' parents, I never took them to fast-food restaurants. They survived and today love vegetables, garden and have even done a little farming, and hold attitudes toward food, health, and justice roughly approximate to mine. And in spite of my aggressive parenting on the subject of food, they are remarkably free of digestive disorders.

Food competency not only takes root in the school and the home but is nurtured in the community as well. While I have made it reasonably clear throughout this book that I do not believe that community and urban agriculture make enormous contributions to food security, I do believe that they offer admirable building blocks for community development and vital training grounds for competent food citizens. These uses should be carefully cultivated and enhanced at every opportunity. A neighborhood that gardens together is, by all measures known to us, a healthier and more desirable place to live. Schools, community centers, and churches that garden together are always stronger than ones that don't. And the more soil that trickles through a person's fingers, the more we can predict that he or she will take care of his or her neighbor, himself or herself, and the land.

Food competency percolates up from the grass roots to city hall, the statehouse, and Capitol Hill. People who are smart about their food choices are also smart and engaged food citizens. They inform their elected officials about important food system policy issues and hold those officials accountable for their actions. They are advocates for food democracy, food justice, and sustainability. To cultivate this ultimate form of food competency, we must expand and extend the world of local and state food policy, which when done right sets a large community table for all to contribute to and feast from.

We have often asked much from our public schools in the past, and now we must ask them to take the lead in fully educating children about food, health, and even agriculture if we are to push back the genuine threat of obesity to our public health. It is imperative that all public school curricula provide adequate time and opportunity for students to fully develop their skills as food buyers, eaters, and preparers and as voters. This must be supported by sufficient federal funding and local support for community gardening and urban agriculture. Land, skill, organizing, and funding are among the ingredients needed to produce fruitful gardens and strong communities. Everyone who wants to garden should be able to do so. Just as we are entitled to clean air and water, we should be entitled to a small spot of earth into which we can drop our seeds and direct our attention. And to give food-competent citizens a place where their knowledge can be most effective, we must make local and state food policy and regional food planning as common as any other feature of civic life. Food policy councils should be a regular part of local and state government. Food systems should be a routine part of the planning profession.

RACE, CLASS, AND PRIVILEGE: MAKE WAY FOR THE NEXT WAVE

The fact that our food system is racist, classist, and sexist should come as no surprise to anyone. When the marketplace fails our communities, and when government fails to intervene effectively and compassionately, people of color, low-income households, and women are the first to suffer. The people who do the dirtiest and toughest jobs in our food system come from the same categories as the people who have the fewest choices, live the shortest lives, and struggle the most to put food on the table. Perhaps the only place in our food system where these folks are not well represented is in the leadership of the groups that are helping them. Rarely will you find a low-income person in a leadership or upper-level management position at an organization committed to promoting food security. And only recently will you find people of color in leadership and professional positions within these same organizations.

Who is affected most by the food gap and who participates in the efforts to narrow that gap are critical questions that deserve much more than a few paragraphs at the end of this book. Indeed, they deserve a book unto themselves. But it is a book whose story is only now unfolding and will best be told five or ten years hence. As a person of privilege and power whose professional agenda for thirty-five years has been to reduce the ill effects of the food system on people who bear little resemblance to myself, I have been intensely aware of both what I can and what I cannot do. There are few examples in the social movement literature, for instance, of one class of people bringing about substantive changes for another class of people. Yes, white people participated in and made a difference in the civil rights movement, as did men in the woman suffrage and women's rights movements, but the victories secured by those movements were due to the leadership of the people most affected by their outcomes—blacks and women. As with these movements, the struggle for equity, access, affordability, healthy food, and food security will ultimately be won by those with the most at stake.

I do not feel that these facts negate my involvement or those of other privileged people in the long-running campaign to end hunger and poverty. Yes, I am privileged. Yet I have chosen to regard that privilege as a gift that I will share as best I can until it loses value or is no longer needed. And as I use the talents God gave me—carefully honed as they were by education, opportunity, and an upper-middle-class upbringing—to make the lives of others at least a little better, I will pave the way for, make way for, and get out of the way of those whose voices more genuinely call out for change than mine ever could.

To ensure that people of color, women, and lower-income people lead the movement to close the food gap, I urge public agencies and especially my own comrades in the food movement to use all means necessary to diversify their leadership and management positions and create a program of leadership development for these people. At the same time, policy-related efforts to change the food system and close the food gap must be more inclusive of those whose lives we are attempting to improve. And while it may go without saying, the growing number of people who are benefiting from a food system that is reaching new heights of health, sustainability, and local connections must unequivocally commit themselves to a food system that is not just profitable and sustainable but also socially just. They can pursue that goal through corporate policies, public policies, faith communities, and individual and organizational acts of charity that give everyone a seat at the table. But most importantly, they must share the same meal.

IT STARTS AT HOME, WHEREVER THAT MIGHT BE

It's early spring in Santa Fe. I've been planting vegetable seedlings indoors in anticipation of moving them soon to our compost-enriched and drip-irrigated garden. This New England gardener has made considerable progress in adapting to the dry, hard clay soils of the Southwest. I'm now more efficient in using the trickle of water that is so precious here. Some would say that I've also mastered the use of shade cloth to tame the effects of New Mexico's brazen sun on my brassicas and to moderate the high desert's nighttime chill that descends on my tomatoes. Yet I'm still baffled by some vegetable-devouring bugs that I never encountered in Connecticut, to say nothing of some cowardly little creature that steals whole plants under cover of darkness, never to be seen in the light of day.

What seasonal produce we don't grow in our garden we buy from the world famous Santa Fe Farmers Market. This is a wonderful institution that nevertheless must do as good a job of serving the city's low-income folks as it does of serving the area's renowned foodies. I will soon attend a meeting of the New Mexico Food and Agriculture Policy Council, which I advised for a while until the members got so adept that they didn't need me anymore. I'll also go to the monthly meeting of the New Mexico Task Force to End Hunger a group that has the unenviable mission of attacking New Mexico's standing as the hungriest state in the country. Because I won't shut up about the value of food policy councils, I've been "drafted" to help develop such a body for the City of Santa Fe. I am proud to do that for a place that has recently enacted one of the most progressive living-wage ordinances in the country. And just because I wanted to do something that I could never have done in Connecticut, I'm serving on the board of directors of an organization that is promoting grass-fed livestock and meat. As the only board member who is not a rancher, I'm also the only person who doesn't show up at meetings wearing a cowboy hat and boots. They seem to like me anyway.

No matter where you live, there are plenty of opportunities to make a difference. That is what is so enjoyable and interesting about food system work. There are hunger and poverty initiatives, local and state food policy campaigns, local food and farming endeavors—to say nothing of tending your own garden—all of which offer anyone with a modicum of interest the opportunity to contribute in multiple ways. But it's important to remember that because the food system is so diverse and complex, it has many interconnected parts, none of which can be ignored for too long before the system falls out of balance. Focus too intently on hunger, and you'll lose sight of its cause. Devote yourself too narrowly to agriculture, and you'll forget about the consumer. Care too much about your own food, and you'll forsake food justice. There are larger purposes in life when all our interests come together. Closing the food gap is one of them.

A NOTE ON SOURCES

Most of the source material for *Closing the Food Gap* is from my childhood and adult experiences with eating, food organizations, and food issues. My professional experience in that regard extends from 1970 to the present and includes significant periods of time in Maine, Massachusetts, Connecticut, and New Mexico. In addition to my recollections of people, events, and conversations I had with colleagues, I have drawn extensively on notes, reports, and articles prepared during the course of my work at the Hartford Food System from 1978 to 2003 and during my time as a Food and Society Policy Fellow at the W. K. Kellogg Foundation from 2002 to 2004. I am particularly grateful to Jack Hale, Hugh Joseph, Gloria McAdam, Jerry Jones, Zy Weinberg, Joel Berg, Maritza Wellington-Owens, and Thomas Forster for granting me interviews in 2006 and 2007 as part of my research for this book.

INTRODUCTION: I'VE COME TO . . . SHOP?

Information about retail food space is from the Food Marketing Policy Center at the University of Connecticut and its director, Dr. Ronald Cotterill.

Dr. Katherine Clancy's essay "Sustainable Agriculture and Domestic Hunger: Rethinking a Link Between Production and Consumption" appeared in *Food for the Future*, edited by Patricia Allen (1993).

Dr. Janet Poppendieck's information is from *Breadlines Knee Deep in Wheat: Food Assistance in the Great Depression* (1986).

CHAPTER ONE: SUBURBIA, ENVIRONMENTALISM, AND THE EARLY GURGLINGS OF THE FOOD MOVEMENT

The original report by Catherine Lerza, "A Strategy to Reduce the Cost of Food for Hartford Residents," was submitted to the Hartford Court of Common Council (the city council) in 1978.

Kimberly Morland's research on access to food stores and the consumption of fresh fruits and vegetables has appeared in the *American Journal of Preventive Medicine* (vol. 22, issue 1, January 2002) as well as in other journals and reports.

CHAPTER TWO: REAGAN, HUNGER, AND THE RISE OF FOOD BANKS

Quotes and comments from Lola Elliot-Hugh and Mark Patton are taken from the Fall 1982 issue of *Seedling*, the newsletter of the Hartford Food System.

The policy report of the Hartford Community Childhood Hunger Identification Project (CCHIP) was prepared and issued by the Hispanic Health Council in 1990.

CHAPTER THREE: FARMERS' MARKETS

Information about Los Angeles is from "Seeds of Change," a study performed by graduate students at UCLA's Graduate School of Architecture and Urban Planning in 1993. Copies can be obtained from the Community Food Security Coalition at www.foodsecurity.org.

California farmers' market information is from "Hot Peppers and Parking Lot Peaches," prepared by Andy Fisher for the Community Food Security Coalition in 1999.

Farmers' market research is from Project for Public Spaces, www.pps.org.

Jon Carroll's piece on farmers' markets appeared in the *San Francisco Chronicle* on June 5, 2002.

Comments from Peter Mann of World Hunger Year appeared as part of a statement made by him and posted to the Community Food Security Coalition website in 2001.

CHAPTER FOUR: COMMUNITY GARDENS

Historical information about community gardens is from an article by Lexia Stoia titled "America Community Gardens," which appeared in the January/February 2007 issue of *Touch the Soil* magazine.

Urban agriculture research is from "Farming Inside Cities: Entrepreneurial Urban Agriculture in the United States," prepared by Jerome Kaufman and Martin Bailkey at the University of Wisconsin in 2000.

CHAPTER FIVE: FOOD BANKS

"Food Assistance Through 'Surplus' Food" appeared in the journal *Agriculture and Human Values* (vol. 22, no. 2, June 2005).

Information on Wal-Mart in New Mexico is from a 2004 interview with Lou Cimalore, food stamp director for Valencia County.

Additional information about food banks comes from "Building the Bridge," a report published jointly by World Hunger Year and the Community Food Security Coalition in 2005.

CHAPTER SIX: RE-STORING AMERICA'S FOOD DESERTS

Information about Hartford's supermarket challenges comes from the report "Connecticut Supermarkets: Can New Strategies Address the Geographic Gaps?" issued by the Hartford Food System in 2006.

Public Voice for Food and Health Policy (no longer in operation) was a Washington-based food advocacy organization that published several reports on food access, including "Higher Prices, Fewer Choices: Shopping for Food in Rural America" (1990) and "No Place to Shop: Challenges and Opportunities Facing the Development of Supermarkets in Urban America" (1996). Several other studies documenting the problem have been published since then, including a report on rural food issues by the New Mexico Food and Agriculture Policy Council titled "Closing New Mexico's Rural Food Gap" (2006).

The quote from former secretary of agriculture Dan Glickman and other related issues is from "Conference on Access to Food, September 18–19, 1995: Report of the Proceedings."

Kimberly Morland's research appeared in the *American Journal of Preventive Medicine* (vol. 22, issue 1, January 2002).

Research on food access and its impact on low-income communities is from the USDA's Economic Research Service, as published in *Food Review* (vol. 24, no. 1).

The Hartford Food System issued a report on the findings of Ph.D. candidate Katie S. Martin titled "Food Security and Community: Putting the Pieces Together" in 2001.

Information about Philadelphia and the Pennsylvania Fresh Food Financing Initiative is from the Food Trust's 2001 report "The Need for Supermarkets in Philadelphia" and its website, www.thefoodtrust.org.

Additional information about Hartford's retail food environment is from the Summer 2002 issue of *Seedling*, the newsletter of the Hartford Food System.

CHAPTER SEVEN: GROWING OBESE AND DIABETIC; GOING LOCAL AND ORGANIC

The results of the Hartford healthy food survey were reported in the Summer 2001 issue of *Seedling*, the newsletter of the Hartford Food System.

Dr. K. M. Venkat Narayan, an epidemiologist for the Centers for Disease Control and Prevention, and Dr. Kevin McKinney, director of the adult clinical endocrinological unit at the University of Texas Medical Center in Galveston, as quoted in an Associated Press story on June 16, 2003, by Janet McConnaughey.

The report from Children's Hospital Boston appeared in the January 2004 issue of *Pediatrics* and was reported on by the Associated Press on January 5, 2004.

The New York City Health Department's Health and Nutrition Examination Survey included significant findings on diabetes levels in New York City and was reported on by *Newsday* on January 31, 2007.

The *New York Times* article on diabetes appeared on January 10, 2006.

The quote about a two-tiered marketplace is from Janny Scott's article "Cities Shed Middle Class, and Are Richer and Poorer for It," which appeared in the *New York Times* on July 23, 2006.

Mari Gallagher's research was published as "Examining the Impact of Food Deserts on Public Health in Chicago" in 2006. For more information, go to www.lasallebank.com/about/july182006_chicagoil.html.

Research by Dr. Rafael Pérez-Escamilla on diabetes in Hartford's Hispanic community was reported by the *Hartford Courant* on March 9, 2004.

The Natural Resources Defense Council's findings on Alar were first reported on *60 Minutes* in February 1989.

The Hartman Group data was reported on May 17, 2006, and January 24, 2007, in *HartBeat*, the firm's online newsletter.

The *Economist* article "Voting with Your Trolley" appeared on December 7, 2006.

The Michael Pollan/John Mackey exchange was reported in the *New York Times* on July 15, 2006.

Information on the Organic Food Production Act is from Clancy's "Sustainable Agriculture and Domestic Hunger" essay.

John Mackey's quote appeared in the *Wall Street Journal* on December 4, 2006.

The *New York Times* article on families escaping from New York City, "Going Up the Country, But Keeping All the Toys" by Alex Williams, appeared on July 18, 2004.

CHAPTER NINE: PUBLIC POLICY

Jerry Kaufman and Kami Pothukuchi's article "The Food System: A Stranger to the Planning Field" appeared in the *Journal of the American Planning Association* (Spring 2000, vol. 66, no. 2).

"Toward Food Security for Connecticut" was prepared under the auspices of the Connecticut legislature's Planning and Development Committee in 1996.

Food system planning comes from the Capitol Region Council of Governments' report "Achieving the Balance: A Plan of Conservation and Development for the Capitol Region" which was submitted to its membership on May 28, 2003.

The discussion of social capital, community development, and participation is drawn from Briggs and Mueller's 1997 report "From Neighborhoods to Community: Evidence on the Social Effects of CDCs" and Martin's report "Food Security and Community."

CHAPTER TEN: INCOME DISPARITIES, POVERTY, AND THE FOOD GAP

The *New York Times* story "Inefficiencies Curb U.S. Aid to the Hungry, Report Finds," March 22, 2007, raised the issue of U.S. hunger relief policies.

Both *Time*'s cover story "Forget Organic. Eat Local" (March 12, 2007) and Michael Pollan's article "Unhappy Meals" in the *New York Times Magazine* (January 28, 2007) discuss the growth and trends in local and organic food.

An article on Whole Foods ran in the February 18, 2004, issue of the *New York Times*.

Survey data from March 23–27, 2003, ran in a story titled "Quality and Prices of Supermarkets" in *Checkbook* magazine, a publication that provides regionalized consumer price and rating services.

Keith Collins's remarks were reported in a USDA press release dated March 7, 2007.

An article reporting on income disparities in the United States appeared in the *New York Times Magazine* on December 10, 2006.

The McClatchy Company story ran in several newspapers, including the *New Mexican* on March 4, 2007.

The *New York Times* ran a story on high-end spending patterns on February 25, 2007, and an editorial on uninsured children on March 12, 2007.

Harry J. Holzer was quoted in "Childhood Poverty Is Found to Portend High Adult Costs," which appeared in the *New York Times* on January 25, 2007.